Newgrange
and the
New Science

BY

Kieran Comerford BE, MBA

Also by the same author

The Christian's Dilemma

R&D and Licensing

Beyond Boundaries

Title: Newgrange and the New Science
Exploring the subtle energies of Ireland's ancient neolithic monument

Author: Kieran Comerford

www.kierancomerford-artandbooks.com

www.magneticloophealing.com

Revised 2011

Revised 2024

ISBN 978-1-291-14290-7

Publisher www.lulu.com

Cover design by Philip Comerford
Cover images from Newgrange.com.

CONTENTS

ACKNOWLEDGEMENTS

I am grateful for the assistance received by a number of people. My thanks to Hugh Kearns who encouraged my interest in Newgrange but who did not live to see his own plans completed and to Mary Armstrong who gave me some valuable inputs at the beginning of this project. My thanks also to Sean McNulty and Kevin Maguire of Dolmen Associates and to my son Philip for assistance with the graphics, to my wife Lynda for her encouragement and editing and to my son Mark for his input on aspects of physics. A number of people read earlier versions of the manuscript and gave me useful comments and pointed out errors. My thanks go to Vincent and Angela Murphy, to Mark Comerford and to Sean McNulty. At the later stages Vincent Murphy took considerable care to edit the document and advise me. Paul Spain came to my rescue in getting the manuscript into publishable form and giving me the benefit of his extensive technical skills. Thanks to all.

I want to thank Dr. Roger Nelson of The Global Consciousness Project at Princeton for permission to reproduce his results and to Harry Oldfield for permission to reproduce PIP scans. Others who gave permissions are Eve O'Kelly, Joachim Trettin and Michael Fox. I want to thank the others who put their work on Wikipedia. If I have failed to acknowledge anybody please let me know.

Kieran Comerford

comerfordkieran3@gmail.com

PREFACE TO THE THIRD EDITION

When I wrote this book in 2011, I little thought that it would be still selling in 2024, and that I would be revising it as a result of interest in a new method of healing using magnets and torsion fields. Reading it now in 2024, I am very pleased to see how relevant it still is. Truth is eternal and does not need to be revised. Nevertheless there are a few things that I feel obliged to revisit in the light of further information or events.

In Chapter Ten, I stated that forms of subtle energy could not be measured by conventional instruments. However, I have since found out how to measure the spiral energy of torsion fields and this method can be seen on my website www.magneticloophealing.com.

In Chapter Sixteen, I addressed the subject of energy from magnets and I described the work of the Irish company, Steorn. I still believe that this company's demonstration of free energy was genuine, but regrettably, they were not successful in commercialising it.

In Chapter Fourteen, I discussed the level of harmful electromagnetic radiation that we are being subjected to. Regrettably, this has continued to increase with Bluetooth and 5G being added in an increasingly energised atmosphere. The corresponding rate of disturbance in adults and children continues to grow.

In Chapter Seventeen, I have described various methods of healing using subtle energy and other techniques. Unfortunately, very few of these have become accepted by the medical and scientific community, but I am glad to say that there is a more general acceptance of meditation in society.

I hope that the book will continue to attract interest and thank you for buying it. I can contacted at comerfordkieran3@gmail.com and would be glad to receive your comments.

Kieran Comerford

December 2024

Chapter One

The Mysteries of Neolithic Ireland

The origins of early settlements in Ireland go back into the mists of time and are shrouded by myth and legend. Definite archaeological evidence from the North-West of the country shows early signs of settlement particularly in County Sligo. One stone construction at Croaghaun in the Ox Mountains has been investigated and found to date from 5,600 B.C.

Many of the ancient sites in Ireland have been described as burial places. Archaeologists have found human remains in these places and so have concluded that they were tombs. However, I have felt that the people living in these places might have had better uses for them than just for honouring their dead. They might have been more intuitive than us since they lived simpler lives closer to Nature. I believe that they might have been sensitive to subtle energies with which we have lost touch and which are not recognised by science. Let me show you what I mean.

Point the index fingers of each of your hands towards each other and bring them together so that the tips of your fingers are about a third of an inch apart (1 cm.). Do this against a dark background. You should see a faint light grey or violet stream of energy going between your fingers. What you have done is to focus the subtle energy of your body onto a point (your fingertips) where it becomes easier to see. Now move your fingers to and fro relative to each other and watch the energy stream follow the movement. I believe that understanding this subtle energy is the key to understanding the purpose of the ancient sites in Ireland.

Neolithic Ireland has fascinated me for a long time. The word "Neolithic" means New Stone Age and it is considered to have begun between 6,000 and 4,000 B.C. It is generally associated with the change from the hunter-gatherer type of existence to one based on primitive agriculture and which brought with it settled communities and the construction of stone buildings but without any visible forms of defence – the mound at Newgrange in County Meath being an example. The investigation of the nature of this peaceful society has occupied me for a number of years and has eventually led to the publication of this book.

Many of the important stone monuments known today in Ireland have been placed on the tops of hills. However, it must be remembered that in Neolithic times the climate this far north was much warmer than it is today and would have been the equivalent to that which we associate with France. The country then was a very different place with much of the lowlands covered in thick forest. This made travelling across land quite difficult and, as in many tropical countries today, it was easier to travel along rivers and so, much of the travelling was done by boat. It is not surprising therefore that an area of County Sligo called the Coolera Peninsula which is surrounded by sea and rivers on three sides should become a focal point for human habitation.

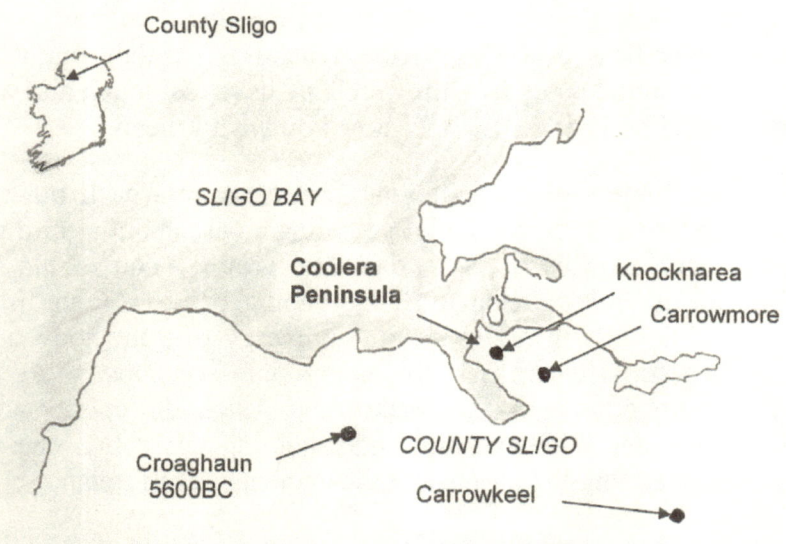

Figure 1. County Sligo and the Coolera Peninsula

When I started to investigate this subject I studied a detailed map of the Coolera Peninsula and counted 128 marked archaeological sites in an area approximately nine kilometres by six. Best known among these is the Carrowmore Megalithic Cemetery, some of which dates from 5,000 B.C. This area covering about one square kilometre is originally thought to have contained over 200 stone monuments but due to quarrying and land clearance over the centuries only about 30 of them have survived. The area has now been preserved and is open to visitors. Excellent guided tours are laid on and a useful publication by Göran Burenhult, one of the Swedish archaeologists who carried out excavations at Carrowmore between 1977 and 1982, can be purchased.[1]

However, the Coolera Peninsula was not known particularly as an area where agriculture developed because there was obviously an abundant supply of seafood to feed the inhabitants as evidenced by the deposits of shells of oysters and mussels found at Culleenmore, Culleenduff, and Grange West. These "kitchen middens" as they are known to archaeologists contain 25 tonnes or more of sea shells and seem to be the dumping grounds of an organised society which had developed the concept of municipal dumps. "Midden" is an old English word for a rubbish dump and many have been found at archaeological sites around the world. In these places food remains such as shellfish and animal bones, ash and charcoal from fires, and broken or worn out tools are found but usually in much smaller quantities than at Coolera.

Figure 2.
The Dolmen at Carrowmore (Tomb 51, *Listoghil*) with its 30 tonne capstone and surrounding cairn material following excavation. It dates from about 3,500 BC.
(photo K. Comerford)

The main part of the site contains a central mound or cairn under which was found a huge dolmen with a capstone weighing 30 tonnes. The mound has been excavated and the stones placed in wire baskets or gabions so that the area around the dolmen can be inspected. This is illustrated in Figure 2 above.

Traditionally, all or most archaeological sites were labelled as tombs. Burenhult states:

> Ireland exhibits a wide variety of megalithic structures, and there are some 1500 recorded monuments on the island. Traditionally, these have been broadly classified into four main types, court tombs, portal tombs, passage tombs and wedge tombs.[2]

These monuments have been seen as tombs because human remains have been found in most of them. It is clear that some of them contained large deposits of bones and may have been ceremonial burial grounds or monuments to the dead. But many of the larger structures have been found to contain only limited amounts of human remains in contrast to the smaller structures. In fact it could be said that the larger the structure the fewer bones it contained. In this book, I set out some alternative theories as to the use of the larger monuments and I ask the reader to bear with me if I depart from the traditional descriptions of them as tombs and instead use the more general name of monuments, mounds or cairns.

The monuments at Carrowmore are laid out in an oval ring around 'Tomb 51' which I will rename Mound 51. These smaller surrounding structures consist of stone circles, passage tombs and dolmens, some containing large quantities of cremated bones. Many of these structures have entrances which point towards the higher central area on which Mound 51 is situated. This mound was partially excavated by the Swedish archaeologists in 1996 and it was found that before its construction, the topsoil had been cleared away and replaced by a layer of rocks and stones. Heavily burnt areas and artefacts were also found.

A cairn of stones and a circle of kerbstones 32 metres in diameter had been built surrounding the dolmen at a later date and human cremations were found close to some of these stones. The dolmen appears not to have been a tomb although, as mentioned above, it did have some burials associated with the outer limits of the mound. If it was not a tomb, what then was its purpose? How and why would a so-called primitive people lift a 30 tonne capstone to make this dolmen? We can only imagine the level of organisation that was involved. I contend that these people were far from primitive and we have to admit that the more we find out about them, the more we realise how little we know about them or their way of life.

To begin to explore this subject we need to understand something about the evolution of Stone Age monuments and to set the developments in Ireland into the broader context of what was happening in other parts of the world at that time. Archaeologists had always assumed that the construction of stone monuments originated with the pyramids of Egypt and spread along the Atlantic coast to Ireland. However in more recent times the results of radiocarbon dating showed that many of the Irish sites are much older than had been thought previously. The stone rows of Carnac in Brittany and the earlier monuments of Carrowmore were shown to be 2,000 years older than the pyramids of Egypt according to Burenhult.[3] The origins of Stone Age monuments and the reasons for their construction is admitted by archaeologists to be a mystery as Burenhult goes on to state,

> The meaning and function of the monuments still remain one of the great enigmas of archaeology, and only recently have we begun to unveil the sociological, psychological and religious background that gave rise to them.

It seems that as far as Ireland is concerned the earliest monuments appeared in the North West in about 5,600 B.C. and spread across the country in a south-easterly direction. Twenty two kilometres south-east of Carrowmore lies Carrowkeel in the Bricklieve mountains, another "Megalithic Cemetery" consisting of 14 passage "tombs" or mounds. These have passageways made from limestone slabs and topped with a mound of stones. It is part of the larger complex of

Carrowkeel/Keshcorran consisting of a large number of archaeological sites including a village with no apparent defences. Unlike Carrowmore, this complex is not 'managed' for tourism and is still in a fairly wild state. One of the structures, Cairn G, which contained some cremated bones, has a roof-box or opening. This allows the sun to shine into the chamber for a month around the time of the summer solstice. The site is believed to date from about 3,000 B.C. It was partially excavated in 1911 in an unprofessional manner and with the use of dynamite.

Figure 3. Queen Maeve's Cairn, 70,000 tons of stone atop Knocknarea, Co. Sligo. *(photo K. Comerford)*

A number of the cairns at Carrowkeel are aligned in a north-westerly direction and in a direct line of site with another monument. This is Queen Maeve's Cairn, a large unexcavated mound on the top of Knocnarea a 327m. hill overlooking Carrowmore and about four kilometres to the west of it. Queen Maeve's Cairn is considered to be the grave of the warrior queen of Irish mythology, Queen Maeve of Connacht. The mystery is, why the people of the time would go to the trouble of quarrying 70,000 tonnes of stones and carting them to the top of the mountain. The huge size of the mound can be gauged by comparing it with the people at its base as seen in Figure 3 above.

Looking in the opposite direction and in a direct line of sight south-east from Carrowkeel in the Bricklieves one comes to the next major archaeological site Loughcrew, 90 kilometres from Carrowkeel. It is located in the central plain of Ireland and has the largest complex of passage mounds in Ireland. Twenty five sites have been recorded stretching for 4 kilometres on a series of hills at the western border of County Meath. Many of the sites consist of substantial cairns, the largest being 45 metres in diameter. The cairns are designated from A to Y and many of them have chambers with astronomical alignments. Some of the cairns contain huge quantities of stones and some of the chambers have had their roofs and covering stones removed. It is clear that many of the stones were used for local building works such as the large stone wall which borders the site. Polished stone spheres of various sizes up to 8 cms. diameter were found in some of the cairns. Similar spheres were found in Carrowmore. What was their purpose?

As we move across the country from County Sligo through Counties Leitrim and Cavan and into County Meath, the incidence of astronomical alignments and engraved stones increases. Carvings in Cairn T at Loughcrew show spiral motifs and a stone showing what could be representations of the Sun. This cairn is aligned with the sunrise at the equinoxes.

Figure 4. Cairn T at Loughcrew. A carved stone showing images of the sun faces east.
(photo K. Comerford)

Forty five kilometres further east into County Meath lies the Boyne Valley complex of mounds of which Newgrange is the most famous. The

7

other two large mounds are called Knowth and Dowth and there are numerous smaller sites in the area mostly unexcavated. Like the other sites they contain wonderful examples of stone carvings of spiral and other designs possibly for astronomical observation or the construction of calendars. They also contained amounts of bones which had been cremated outside and brought inside for some ritual purpose. The official brochure of the National Monuments and Historical Properties Service of the Irish Government describes these sites as "truly remarkable remnants of a highly evolved society".

Newgrange consists of a passage leading to a cruciform chamber. The passage is constructed of standing stones roofed over with flat slabs and covered with a huge mound containing over 200,000 tonnes of earth and stones. The outer boundary of the structure is marked by a circle of ninety seven horizontal kerbstones. Many of the kerbstones have intricate designs carved on them. Above the entrance is a roof-box which allows the sun to shine into the passage. The structure is aligned so that on the morning of the winter solstice on 21st December each year the beam of rising sun reaches into the inner cruciform chamber. The entrance is surrounded by tonnes of quartz found on the site and built into an impressive wall during its excavation and partial reconstruction.

It is now accepted that Newgrange had an astronomical purpose. Perhaps it was to mark the exact mid-winter day and to function as a calendar so that crops could be planted and rituals observed. However, in 1964, two archaeologists, O'Riordan and Daniel referred to theories based on the solar alignment as a "jumble of nonsense and wishful thinking".[4] Archaeological thought and understanding have now moved on and the principle of astronomical alignment is now well established.

Newgrange dates from 3,200 B.C. but Knowth, one kilometre to the west, is possibly earlier and is surrounded by seventeen smaller satellite mounds. The map in Figure 6 shows the main sites mentioned so far and indicates the direction of evolution of these structures starting at Carrowmore in 3,550 B.C. and culminating in the magnificence of Newgrange.

Figure 5. Newgrange Co. Meath, 200,000 tonnes of earth and stones. What was its real purpose? *(photo reproduced with permission from Newgrange.com*

All these sites have visual contact with the next site in line and many of them could have been in use simultaneously. Thus it is possible that there may have been direct communication between them by the use of signal fires. I believe that there is a connection between these monuments and that it indicates a purpose other than merely honouring the dead.

My interest in this subject started with the study of meditation and Eastern philosophy in the 1980s. I was intrigued by the Eastern energy concepts of chi and prana which are not understood or recognised by Western science, and I went on to study the parallels between Eastern philosophy and modern physics. Here my training as an electrical engineer helped me as I started investigating the subject of subtle energies in general. A visit to a conference in New York led me to start thinking about the possible functions of Newgrange other than that of a tomb or astronomical observatory, and I came to the conclusion that there was some form of technology involving subtle energy which we did not understand but which was known and used by the Newgrange inhabitants.

Figure 6. Stone Age technology evolved from the area around Carrowmore and moved south-east (as shown by the large arrow) to Carrowkeel, Loughcrew and finally Newgrange.

By this time I had moved in my professional life to work in a government research institute where I specialised in the management of research and development. I eventually set up my own consultancy business in R&D management and this gave me more freedom to pursue my interest in Newgrange and its energy. One day it occurred to me that Newgrange was the final stage of a huge R&D project that had started many centuries earlier on the north-west coast. I set out to investigate the evolution of these structures and I now believe that the structures north-west of Newgrange represent earlier models built to understand and develop a Stone Age technology based on subtle energy.

What was this subtle energy and what was the technology used for? Why did people who apparently had no earthmoving machinery or cranes shift huge quantities of stones and earth to build these stone monuments? Were they more than monuments? Did they have an everyday practical use and if so what was it?

When I started my investigations into the Newgrange technology I did not know the answers to these questions, but as I worked, various ideas came to me and I found myself looking into the whole subject of energy fields. I found that important new developments are taking place but outside mainstream science. The acceptance of these developments requires a radical new approach to science, a major shift in consciousness. The following chapters chronicle my investigations and lead to a whole new way of looking at science and how it can solve many of the present problems in the world.

Chapter Two

From Partnership to Domination

The seeds of my research project on Newgrange were sown on the 18th of May 2001 when I found myself sitting in the Synod Hall at the Cathedral of St. John the Divine on Amsterdam Avenue in New York City. I was there to attend The Prophets' Conference, a conference run at regular intervals which brings together luminaries and visionaries who share their wisdom with others and discuss how the world can be changed for the better. The speakers came from a wide variety of backgrounds and included scientists, philosophers, psychologists, spiritual leaders and shamans. Most of them had one thing in common; they had broken away from mainstream thinking and were giving out a new message, one free from traditional beliefs.

One of the speakers was an Austrian lady named Riane Eisler. She has won recognition for her work as a social historian and for her book *The Chalice & The Blade*.[1] Her early childhood where she witnessed Nazi persecution in Austria left her with an insatiable desire to understand the reasons for man's inhumanity to man (and to woman). In her book she proposes the thesis that society has two possible methods of organisation which she calls the "dominator" model and the "partnership" model.

She explains that we have chosen the dominator model. The society we live in is organised on the basis of power and domination. This has evolved over time. In the distant past when there were fewer people in the world, there was an abundance of food and less need for shelter, especially in regions of favourable climate. People were able to live in peace and harmony and they shared their resources equally. Since there were no fears of shortages, it was not necessary to hoard. The concept of building up a store of wealth did not exist. As the population of the world increased, more marginal regions of the world were populated and when changes in climate occurred, shortages of food and shelter became apparent. Competition for scarce resources led to conflict and those who were physically strong gained at the expense of those who were weak. The strongest fighters became chieftains and amassed wealth and power. They then annexed neighbouring lands and eventually became kings. They used their wealth and power to exploit others, made their own laws and appointed their followers to enforce their decrees. They formed armies and created a culture which until relatively recently expected most men to go through some form of military training. Men were conditioned to use their superior physical strength and power over other men and also over women.

When kings adopted or converted to a new religion they used their armies to ensure that their subjects followed suit. When they conquered a new country, those who did not convert were put to death. Their power was also passed on to church leaders who supported them. When the Emperor Constantine converted to Christianity during the fourth century A.D., Christianity became the approved religion and Christian bishops who had previously been persecuted, were now able to exert power over those who did not follow their beliefs. Apparently this dominator society had become so well established that the bishops were able to overlook the well documented facts that Constantine ordered the execution of his son Crispus on an unsubstantiated charge of conspiracy and had his second wife Fausta killed by imprisoning her in an over-heated steam room.[2]

The temporal power of the Christian bishops steadily increased, leading to the suppression of ideas which opposed the development of the hierarchical system of power. Elaine Pagels, Professor of Religion at Princeton states in her book *The Gnostic Gospels* that other less 'orthodox' forms of Christianity were suppressed and those who followed them were declared heretics.[3] These less orthodox forms were based on the principle of "gnosis" or direct knowledge of God through mysticism. As all Gnostics considered themselves equal there was no need for a hierarchy of priests and bishops. If people could obtain direct knowledge of God they did not need to have others to act as intermediaries.

In a previous book, *The Christian's Dilemma: A guide to the new spirituality,* I explained that reincarnation formed part of the beliefs of the early Christian Church but it was declared a heresy by the Council of Constantinople in 553 A.D.[4] Whether this rejection of reincarnation was ever formally ratified by the Church is admitted to be still an open question by *The Catholic Encyclopaedia*.[5] However, it seemed to be convenient to leave the subject to one side as Church leaders felt that people would become lax about following their religion if they thought they had many lifetimes in which to achieve salvation.

Reincarnation surfaced again in the twelfth century with the Albigensian Heresy. Based around the town of Albi in Southern France, an ascetic group who called themselves the Cathars was set up as a reaction to the sorry state of the Church at that time. Bishops and priests lived in luxury and abused their positions by charging fees for Church services and for granting favours such as approving illegal marriages.[6] The Cathars, who considered themselves to be followers of the original Christian ideals, believed in reincarnation and pursued an austere lifestyle with a semi-vegetarian diet. They gave recognition to women who were allowed to join their clergy. Many of the nobility of the Languedoc region began to join the Cathars and the Church became alarmed at their growing influence. In 1208 Pope Innocent III launched a Crusade against the Cathars and an army set out to suppress this group whom they accused of rejecting the Eucharist, the resurrection of the body and worship of the cross. It is estimated that over fifteen thousand people were put to death in the town of Béziers alone. Many of them were ordinary Christians who were killed because the soldiers could not tell them apart from the Cathars.[7] By 1233 the Inquisition, which was to control heretical outbursts for the next six hundred years, had been set up and the Cathars were finally suppressed in the siege of Montségur in 1244.

The dominator form of thought policing has persisted through the ages. Galileo was persecuted for saying that the Earth was not the centre of the universe but revolved around the sun. This idea had originally been proposed by Copernicus and it was considered that such theories might lead people to think that they did not have an immortal soul and were simply part of nature and not superior to it. In 1633 Galileo was brought before the Inquisition in Rome and was made to renounce all his beliefs and writings supporting the Copernican theory. He was imprisoned and his works remained banned for two hundred years.

For hundreds of years Jews were forced to live in ghettos and were prevented from taking part in guilds or trades. The only business left open to them was money-lending, an activity which was not considered 'moral' for Christians.

Further evidence of the Church's support for the dominator model continued up to the time of the Second World War. For example, in 1939 two German cardinals sent telegrams of support to Hitler after the failed attempt to assassinate him on the 8th of November in that year.[8] This was a time when wives were told to be obedient to their husbands and women who gave birth had to be 'purified' by the questionable practice of churching. Thankfully things have improved, although in present times, many women question why they cannot become priests and take a full part in the running of the Catholic Church.

But it was not always so. Riane Eisler challenges the popularly held notion that early societies were all primitive and less civilised than ours. She draws attention to the numerous mistakes of early archaeologists who interpreted all they saw from their own perspective which she calls "the lens of their own male or dominator consciousness". Queens were assumed to be kings and ritual was confused with warfare. She has made a special study of the Minoan civilisation which existed on the island of Crete from about 6000 to 1500 B.C. She describes the recent discovery of the technologically advanced and socially complex society of Minoan Crete as an "archaeological bombshell". Archaeologists could not understand how the existence of such a highly developed society could have remained undiscovered for so long.[9]

What came to light was amazing. They found vast multi-storeyed palaces, organised cities and districts with road systems, houses with sanitation and magnificent works of art. These portrayed a society which was peace-loving and worshipped a goddess. Nowhere could the archaeologists find indications of fortifications or defences even on the seashore from where one would expect invasions.[10] Their government was based on the sharing of wealth and there were no signs of law enforcement by armed might.[11] The argument for the existence of an egalitarian society was supported by the fact that there were no statues to kings or queens. Nor did their art depict scenes of battles or hunting as found in other later so-called 'civilisations'.

Minoan society recognised the importance of the female aspect as the giver of life. However, it was not a matriarchal society although there is ample evidence of female rulers and goddesses. Men and women were equals and engaged in work, sports and recreation together. They were not repressed in their attitudes to sexual expression and the liberal atmosphere which prevailed contributed to a peaceful and harmonious co-existence of the sexes.

Figure 1. Knossos in Minoan Crete, a highly civilised partnership society
(photo from Wikimedia Commons)

This was a perfect example of Eisler's partnership model. Men and women each had a role to play in society and each respected the contribution of the other. Society functioned on the basis of mutual respect and partnership without the need for domination of any person by any other. This is not to say that it was a Utopia. It was not free from disease and suffering. It also had natural disasters and suffered a number of very destructive earthquakes and volcanic eruptions.[12] Recent research suggests that the Minoan civilisation extended to the island of Santorini in the north which was almost annihilated by a huge volcanic eruption. Measurement of tree rings in bog oak found preserved in Irish bogs suggests that a huge volcanic explosion which spread ash right around the earth, took place in 1629 B.C. Archaeologists have found a chaotic jumble of material along the northern coastline of Crete suggesting that the area was devastated by a tidal wave at least as big as the 2004 tsunami in the Indian Ocean. Radiocarbon dating of this material gives a date of 1600 B.C., thus confirming the most likely date for this event which destroyed the Minoan economy.[13]

We know that the Minoan civilisation eventually succumbed to invaders from Mycenaean Greece. Worship of the goddess figure was gradually replaced by worship and propitiation of the male gods of war and thunder. Worship based on love became worship based on fear. The dominator society eventually became established. Partnership as a model for society was dead.

Archaeologists have traditionally viewed early societies as primitive and warlike but recent discoveries have led some of them to change their minds. Jonathan Haas of Chicago's Field Museum has spent twenty years working to prove the hypothesis that warfare was the stimulus that led to the beginning of what we call civilisation. For more than one hundred thousand years humans had roamed the earth as hunter-gatherers or lived in small villages as farmers. About six thousand years ago something happened which caused some of the small villages to grow into highly organised cities. The question was what? Many archaeologists believe that the need for villagers to defend themselves against attack led to larger defensive settlements and these in turn led to the need for higher levels of organisation and increased specialisation. From this, they believe, sprang the cities of the ancient world in areas such as Egypt, Mesopotamia (Iraq), China and India. These cities had fortifications and built massive monumental structures such as pyramids. However, recent discoveries at Caral, 120 miles north of Lima, Peru reveal a highly civilised society living in a city of pyramids and dating back to 2627 B.C. with no signs of any defensive structures.[14]

For years archaeologists had been looking for what they called a "mother city", one without any later building on the same site. They wanted to study a city in pristine form free from the ravages of later settlement, so

17

that they could investigate its origins and discover how it came to be organised. There they expected to find evidence of warfare and prove the theory that warfare gave birth to civilization. But in Caral they found evidence of a highly civilised and peace loving society which grew cotton and traded textiles and fishing nets with the coastal fishermen who gave them fish in return.

Jonathan Haas and his wife and collaborator Winifred Creamer scoured the area for signs of warfare and found none. They were forced to abandon the warfare theory. In the course of a *Horizon* programme on BBC TV Haas says:

> You seemed to really have the beginnings of that complex society and I'm able to look at it right at the start and I look for the conflict and I look for the warfare, I look for the armies and the fortifications and they're not there. They should be here and they're not and you have to change your whole mind-set about the role of warfare in these societies and so it's demolishing our warfare hypothesis. The warfare hypothesis just doesn't work.[15]

When I read about the Minoan and other early civilisations I began to wonder about such civilisations in Ireland. Numerous facts about early Irish societies began to come to my notice and I set out to look for more. In particular, I was struck by the absence of forts or defensive structures of any kind in Neolithic settlements. One example of these recently discovered is the Céide Fields Neolithic field system in North Mayo in the west of Ireland. The fields, enclosed by stone walls, have been preserved intact by a cover of blanket bog that is more than four metres deep in places. The remarkable thing about Céide Fields is that the people left this area 5,000 years ago when it became too wet to inhabit. Everything was covered by a slowly growing bog which has preserved the landscape intact until this day.

Although the area is too big to investigate by digging, four square miles of it have been mapped out by probing the bog with long metal poles and it is probably even more extensive. Investigations have revealed a series of rectangular fields constructed of 250,000 tonnes of stones built into walls. It is clear that this was the work of a highly organised society. An interpretive centre has been built on the site and the display material refers to a peaceful society of "unprotected family dwellings scattered through the landscape indicating that there was no threat either from within or outside the community".[16] Later societies found it necessary to group their houses together behind common defensive walls. No defensive walls exist at the Céide Fields. It was a simple "countryside of homes scattered through the landscape surrounded by their garden walls".[17]

Figure 2. Céide Fields in County. Mayo Ireland. A Neolithic field system with its interpretive centre in the background.
(photo K. Comerford)

In looking at early civilisations I was struck by the contrast between their buildings and those of later civilisations that were definitely living according to the dominator model. The surviving buildings of early civilisations seemed to have been built for ritual purposes or to have astronomical alignment. Those of later civilisations were erected as fortifications or to honour great leaders. They were built either to impress or to terrify. Those built to impress usually housed tombs of kings or queens and when archaeologists found earlier Neolithic structures it was assumed that they were also tombs.

Many people disagree with the theories of former schools of archaeologists who labelled all of the early structures as tombs simply because they contained bones. In their book *Keeper of Genesis* Robert Bauval and Graham Hancock are critical of the academic archaeologists: "the practice of burying the dead there…was a later adaptation effected by people who were unconnected to the genesis of the site but who sought to be interred in a place that was imbued with prestige and sanctity." [18] They go on to give the example of the burying of mediaeval archbishops under the flagstones of cathedrals. This, they say, cannot be taken to mean that such cathedrals were built as tombs.

But what if many of these early structures were not intended as tombs? What if they had other purposes? What if they were not built for any egotistical purpose, but instead built in partnership to fulfil some function that we can only guess at? What then was the purpose for which Newgrange was built?

It is known that Newgrange was built by a highly civilised society. Stone was transported from the Cooley Peninsula thirty miles to the north and from the Wicklow Mountains thirty miles to the south. There has been a lot of speculation in recent times about the construction of these types of monuments, and attempts have been made to find out how it was done. Part of Stonehenge in England was made from Welsh stone transported over two hundred miles. The stones of Newgrange are not as massive as those of Stonehenge but they still weigh up to ten tonnes. These large stones were quarried locally but still had to be transported some distance and then put in place. It is clear that this construction must have taken place over many years and that a high degree of organisation and vision was involved. This was not the work of a primitive warlike society.

Newgrange is only one of 1,500 megalithic sites in Ireland. Megalithic means "large stones and it is clear that these societies could have built many fortifications in stone if they had needed to but no defensive structures from this time have ever been found. What then was the origin and purpose of the megalithic monument building culture?

Chapter Three

The Spiral

One of the most significant designs on the many stones in and around Newgrange and its neighbouring sites on Knowth and Dowth is the spiral. Sometimes it is a single spiral. Sometimes it is a double or triple spiral design. The spiral is also found engraved on stones at numerous early sites such as Loughcrew. Could the spiral be the key to understanding the real function of these early Neolithic sites? Could this point to an alternative explanation for a type of civilization which is completely different to anything we understand today?

Figure 1. Newgrange
triple spiral design
(photo reproduced with permission from Newgrange.com)

I started to think about the spiral after making a visit to a man who describes himself as an expert on geomancy, living water and subtle energies. He runs an organisation called the Heritage Awareness Group in Dublin which makes study visits to ancient sites. On the wall of his living room was a large model in copper piping of the triple spiral design from Newgrange as shown in Figure 1 above. He told me it contained energised water and invited me to test it. Puzzled, I agreed as he took it down from the wall and placed it on the floor. He then invited me to take off my shoes and stand on it for two minutes. I did this and experienced a kind of tingling feeling and did not think any more of it, but on the way home in the car I noticed that I was buzzing with energy. My wife, Lynda, also picked up the energy which she sensed from me when I came home. That night I was still feeling the buzz and was unable to sleep.

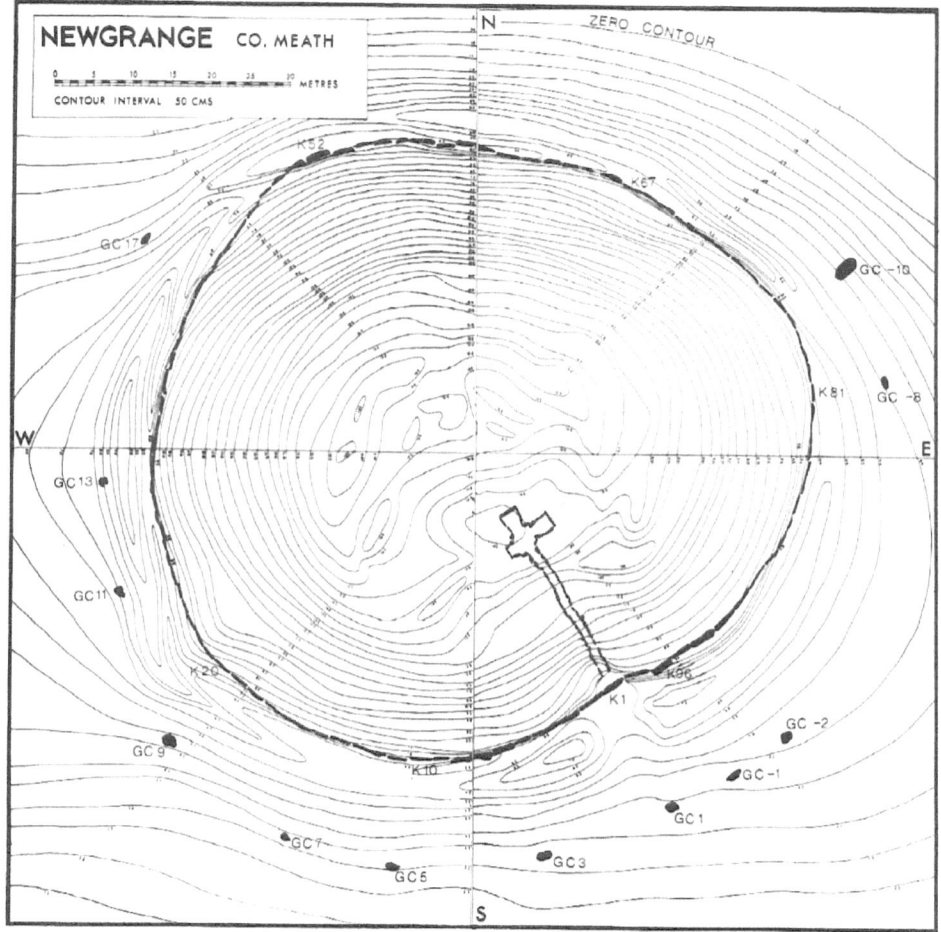

Figure 2. Plan of Newgrange showing the cruciform chamber, 97 kerbstones and 12 remaining standing stones.

(Reproduced with permission from the estate of Michael J O'Kelly)

23

My curiosity led me to join the Heritage Awareness Group[1] a few weeks later and to travel with them on a day trip to the valley of the Boyne. The Boyne river flows from west to east and enters the Irish Sea at Drogheda thirty miles north of Dublin. The valley of the Boyne contains numerous historic sites dating from the Neolithic Age right up to the Christian era and includes the structure at Newgrange. Many of these sites have not yet been excavated.

We started our tour at Newgrange. This is referred to as a "passage tomb" but I will use the term "passage mound" which is less presumptuous as to its original purpose. The mound itself is 11 metres high and up to 85 metres at its widest. As already mentioned, it contains over 200,000 tonnes of stone and earth. The first thing that strikes one on approaching the mound is the kerbstone across the entrance which is covered in spiral designs. On entering the passage one walks to the cruciform chamber which can accommodate about twenty five people. The chamber has a corbelled roof made from flat slabs resting on each other and gradually getting narrower towards the top. It is six metres high. Carvings inside the chamber show spirals and other designs. Once the tour guide has given a description of the chamber, the lights are turned off and a simulation of the winter solstice sunrise takes place. This is done by moving a light across the entrance to the passage. The beam of light can be seen gradually coming up the passage until it reaches the chamber. The effect is magical and highly symbolic. Many explanations are given for its symbolism but the one that struck me most was the suggestion that on the day of the winter solstice, the male sun god penetrates the female earth god to ensure a fertile growing season in the coming spring.[2]

Outside the mound, the leaders of the Heritage Awareness Group told us that the mounds in the valley are all aligned over the crossing points of underground streams and they then led us on an exercise in dowsing or divining the positions of the streams. I was highly sceptical of my ability to dowse but found that, under proper instruction, I could do it just like the others. It was impressive to see a group of twelve people holding divining rods walking in a line radiating from the centre of the mound and all simultaneously detecting an underground stream at the same point, and without any prompting.

Figure 3. Newgrange. Crossing underground streams create a vortex of rising energy *(Image drawn by Dolmen Associates)*

At a few points on the site we were shown the crossing points of streams which were described as vortices. Here, we were told, the movement of the water in two directions causes a spiralling flow of energy which is detected by a pendulum or by dowsing rods. This was my first explanation of a possible meaning of the Newgrange spiral, a point of energy concentration. I did not know what kind of energy it was as it certainly was not the kind of energy which would be measured by conventional instruments. I decided to accept the fact that it was a form of subtle energy and I reaffirmed my determination to learn more about it.

We then visited the mound at Knowth about one kilometre west of Newgrange. This is larger and possibly earlier and is surrounded by seventeen smaller satellite mounds. At the time of this visit the site was still being excavated and so access was limited and it was not possible to go inside any of the passages. The mound is 95 metres in diameter and 12 metres high. Of the 127 kerbstones around the outside many are inscribed with a variety of designs including spirals, waves, and what may be either a sundial or, as argued by others, a complex solar calendar dividing the year into sixteen months.[3] The truly remarkable feature of Knowth is that it has two passages facing east and west and aligned with sunrise and sunset at the time of the equinoxes.

Numerous books have been written suggesting the purpose of the mounds in the Boyne Valley. I have spoken to many people who have their own theories stemming from folklore or their own esoteric studies. The late Hugh Kearns, whose friendship I valued, came up with the theory that Newgrange was the first Son et Lumière in history, the original light show! This is described in his entertaining book *Newgrange: The Mystery of the Chequered Lights.*[4]

His theory was that when the beam of the rising sun entered the chamber through an opening above the entrance called a roof-box, it was reflected back out through the entrance by means of a mirror made of polished gold. The mirror was suspended from a tripod arrangement of poles placed in the three alcoves at the top of the chamber. Moving the mirror from side to side would have created a "chequered lights" or flashing effect for the crowds assembled in the natural amphitheatre on the river bank opposite.

You might wonder, as I did, why the phrase "chequered lights" is used. The ancient name for Newgrange is Brú na Bóinne (Mansion of the Boyne). In an old text called *Colloquy with the Ancients* a resident of Newgrange is asked where he comes from. His reply is, "Out of yonder Brú chequered with many lights".[5] There is also and old Irish word "Brú" meaning a womb, suggesting that the male sun god penetrating the female earth god is the true meaning since as can be seen in Figure 2, the layout of the entrance, passage and chamber has a resemblance to the female reproductive organs.

As well as purposes related to their astronomical alignment it is also suggested by archaeologists and others that the chambers were used by the elders to enter the trance state. If one treats the chambers as resonant cavities it can be shown from their dimensions that they have resonant frequencies lower than the longest organ pipes and similar to the alpha waves in the brain[6]. When the human brain is in the alpha state of 8-10 Hertz (Hz, cycles per second), a person becomes very relaxed and enters into a state of trance or meditation. The total sensory deprivation inside the mounds would, when combined with the shape of the chambers, lead people inside to maintain trances for long periods.

Acoustic archaeology is a relatively new subject. Until recently archaeologists concerned themselves with what they could see and touch in ancient sites, but their attention is now being drawn to what they can hear. In his book *Stone Age Soundtracks: The Acoustic Archaeology of Ancient Sites*,[7] Paul Devereux gives details of acoustic measurements carried out at a number of ancient sites including Newgrange and Stonehenge. All of the sites surveyed were constructed in such a way that sounds emanating from a focal point would be amplified and would carry to the outer reaches of the site. This would have facilitated chanting, speaking and shamanistic rituals involving trance states. Most sites had resonant frequencies in the range of 95-120 Hz, corresponding to a male voice range.

In Newgrange, sound patterns of standing waves, (alternating points of maximum and minimum volume), were measured along the chamber. There were twelve in all, leading to the suggestion that the spiral and other diagrams represented sound-waves. A carved lintel stone over the roof-box has twelve diamond or lozenge shaped designs in a horizontal row and it has been suggested that these are a graphic representation of the standing waves.

As well as investigating sounds in the audible range, Devereux reports on work by a team from Reading University who also investigated lower frequencies or infrasound. They discovered the same type of resonances as mentioned by Devereux above. These were in the range of 4-6 Hz. At Camster Round in Caithness, Scotland, they carried out an experiment in which a drum was struck at four beats per second. The subjective experiences of the audience were then recorded. Aaron Watson, one of the researchers stated: "Although the presence of infrasonics in sufficient quantities could not be confirmed, it was certainly apparent that the tomb interior was conducive to the creation of unusual experiences".[8]

I consulted an expert in folklore who suggested to me that the chieftain, who was the most psychic person in the tribe, used the trance state to go on a vision quest similar to that undertaken by North American Indians. The vision quest was represented by a spiral moving back into the past for the purpose of discerning the truth to settle disputes, or moving forward into the future to seek information such as the best crops to grow that year.[9]

Trance states can be induced by chanting, drumming and by the use of hallucinogenic drugs. The fact that the chambers resonated at vocal frequencies would amplify the chanting as well as resonating with the alpha waves in the brain. The use of hallucinogenic drugs by elders is reported to be a theory highly regarded in academic circles. These would have been derived from plants like the psilocybe mushrooms used by native peoples in Mexico. Scientific research shows that images seen in states induced by certain drugs are very similar to those carved on the stones of Newgrange and Knowth.[10]

The archaeologist Jeremy Dronfield has referred to the "vortex or tunnel experience" commonly encountered in altered states of consciousness and described in near death accounts to explain the appearance of circles and spirals. He says that the passages in Newgrange "were associated with a complex of consciousness-altering practices involving the induction of subjective visual experiences by means of flickering light, hallucinogenic substances and neuropathology". In particular, the passages induced "the visual impression of looking into, or moving through, a vortex or tunnel". [11]

The images seen represent primary forms of organisation in nature. Physicists now acknowledge that everything from DNA to plant and animal cells to hurricanes and tornadoes to galaxies form spirals or vortices. Based on projective geometry, Lawrence Edwards showed how plants and animals grow according to certain natural laws and astronomical cycles. His studies of bud shapes, begun in the1960s and continued for thirty years, built up a comprehensive picture of how plants are drawn upwards by natural forces and follow geometric paths to reach their mature shapes.

Most of his work was done in Argyll in Scotland but towards the final stages he received confirmation of his theories when a collaborator, Graham Calderwood duplicated his results 180 miles away in Aberdeen. His work is described in his book *The Vortex of Life*.[12] In it he shows how natural shapes such as buds, eggs and embryos evolve according to the principles of projective geometry. Plants grow according to path curves and follow spiral or vortex patterns. The best example is seen in the growth of a pine cone as seen in Figure 4 below.

Figure 4. Lawrence Edwards' work shows how plants grow in spiral paths
(photo from Wikipedia commons)

The mystical significance of the spiral is beautifully described by Jill Purce in her book *The Mystic Spiral*.[13]   The spiral is held by many traditions to be the symbol of the connection between the point and the infinite, between the inner and the outer, or between humankind and God. The spiral when considered in three dimensional form curves back on itself to symbolise the evolutionary progress of humans towards the Divine and simultaneously, the involutionary progress of the Divine becoming established in us. This version of the spiral drawn around an imaginary torus or doughnut curves inwards to its centre then outwards around the outside to join up with itself and recommence its continuous inward and outward movement. The spiral represents the duality of our existence and we should really consider both the inward and outward spiral together or, as Jill Purce says:

> The spiral has actually returned by winding onto its source.
> Its 'end' is not a second and therefore relativating infinity
> as implied by the single spiral. The duplication of the One
> is simply *the One looking at itself, and in so doing
> becoming subject and object:* this is the duality by which
> all is known.

This parallels the description of the creation of matter out of consciousness as described in the Vedic literature of ancient India which comes from an oral tradition predating Newgrange and Knowth. "Curving back on my own nature, I create again and again" is how it is described in the *Bhagavad Gita* (Chapter 9, verse 8). Consciousness knows itself and through the process of knowing causes the manifestation of the body. The body, being conscious, knows itself and this sets up a cosmic vibration which goes on indefinitely - the hum or 'OM' of creation. But the Vedic literature goes further to state that the true knowledge of the 'Self' is obtained through reaching a settled state of awareness in meditation. This is represented by the stilling of the mind from the awareness of many thoughts to the awareness only of itself, the collapse of infinity to the point. This is described in verse 1.164.39 of the *Rig Veda*.

> The verses (of the Vedas) exist in the collapse of
> fullness in the transcendental field.
> In which the devas responsible for the cosmos
> reside.
> For (the person) who does not know this field what
> will these verses accomplish? [14]

The reference in the last line is to the idea that the knowledge of the Vedas is only understood at a superficial level by those who do not access the knowledge through meditation. The Vedas were cognised or understood intuitively by ancient seers or "rishis" who lived in caves in the Himalayas before the time of Newgrange. This knowledge, which we are only now beginning to understand, is leading us to mental technologies which can be used for the benefit of everyone.[15] By curving in on ourselves we can get to that place from which we can emerge in a newly created form. Perhaps this knowledge was available to the seers of Newgrange as they sat in meditation in the cruciform chamber.

Perhaps ancient civilisations had access to higher dimensions and to forms of energy which we are only beginning to rediscover. The highly evolved society which built Newgrange was clearly based on partnership rather than on domination. It is clear that there was a complex society with elders who had a "knowledge of exotic items and ideas".[16] The passage mounds only contained remains of a small number of people.[17] This could lead archaeologists to conclude that they were the remains of kings and their families which one would find in a dominator society. It is more likely however, in the light of our present knowledge, that the remains would have formed part of complex rituals of reverence to departed elders as are common in shamanism. Entering the trance state, the living elders would have attempted to link with the spirits of the dead to help them on their passage to the next life. In other rituals they may have linked with spirit guides as part of their vision quest.

None of the many Neolithic settlements in the Boyne Valley contain any signs of defensive constructions. This was a farming society. We are not talking about hunter-gatherers as in the earlier Mesolithic. These people were well capable of building in stone, and if they wished to have strong defences against attackers, that is the material they would have used and such constructions would have survived to this day.[18] Instead, they were defenceless. Perhaps they understood intuitively the message in the Yoga Sutras of Patanjali which states that "in the vicinity of non-violence, all hostile tendencies are lost".[19] Like the yogis of India, they were in touch with a higher energy. We will explore this energy in the following chapters.

Chapter Four

Torsion Fields

In ancient times the spiral was understood on an intuitive level. In more recent times it began to be understood as a physical phenomenon. For example, the smoke ring which is easily seen in nature is a form of three dimensional or toroidal spiral. In 1867 the Belfast-born physicist Lord Kelvin demonstrated that smoke rings were a stable form of matter. With a smoke generator he produced smoke rings which bounced off each other when they collided. They did not intersect each other and it was not possible to cut them with a knife. They just slid away. He described them as vortex rings and propounded the theory that the spinning vortex was a fundamental unit of matter. This was named the vortex atom theory and was espoused by James Clerk Maxwell, famous for his theory of electromagnetism. Unfortunately the vortex theory was forgotten with the excitement of new possibilities introduced by Quantum Theory although physicists may be coming close to it again with a theory of torsion fields which will be described later in this chapter.

A novel approach to the spiral or vortex is described in a book by Callum Coates on the amazing work of Victor Schauberger, an Austrian who lived from 1885 to 1958.[1] Schauberger was a forester who lived close to nature and observed the flow of water in mountain streams and rivers. He produced revolutionary ideas on how water in rivers can be kept oxygenated and stated that rivers should be allowed to flow naturally without being channelled between steep walls. This, he contended, would prevent floods and damage to bridge foundations by scouring. Many of his ideas for directing water flow were adopted.

Unfortunately Schauberger was no diplomat and was not able to converse with scientists and engineers in terms they understood, and so his ideas went unrecognised. He produced many ideas for machines which made use of spiralling liquids and gases. These included a type of flying saucer and a new method of propulsion for boats. He rejected the use of combustion in producing energy because it caused pollution. Instead of using explosions he developed the idea of implosion, where liquids and gases are accelerated by putting them through spiral pipes and vanes.

At the beginning of Chapter 3 in introducing the subject of the spiral, I described how I experienced a buzz of energy after standing on a spiral copper coil of energised water. I was told that this water was imploded water which had been developed following Schauberger's ideas. Schauberger was convinced that the spiral or vortex was a device which raised the frequency of vibration of matter until it entered a higher dimension from which energy could be obtained.

A physicist would ask what Schauberger meant by the "frequency of vibration of matter", stating that it is too imprecise to be meaningful. Spiritual people are fond of telling us that we live in a "very dense vibration" on Earth and that this earthly plane of low vibrational energy is a very difficult place in which to function. They contend that a transition to a higher plane or dimension is experienced at death or, on a more temporary basis, during meditation. They tell us that these non-physical realms operate at a higher vibration and that they are much more pleasant places to inhabit. Non-physical things are difficult to explain in physical terms and this may explain why people operating from an intuitive level have difficulty communicating their ideas to scientists. However the physical aspects of the spiral can be explained. It all comes down to one word "spin" and this concept is increasingly being recognised in various branches of physics.

If matter is put into a spiral or vortex, it will be caused to spin. It will spin more quickly or be accelerated as it moves from the edge to the centre. For example if water is pumped tangentially into the top of a vessel with a hole in the bottom, it will be seen to form a whirlpool. This is what happens when you empty the bath. As long as the water can be taken away fast enough, a whirlpool will form. Other examples of this in nature are hurricanes and tornadoes, where winds can spiral at speeds up to 300 miles per hour. The reason this happens is that the angular momentum of the air is not reduced when it is caused to flow in tighter and tighter circles. The energy must go somewhere so it goes into accelerating the air. This applies to any substance. The classic example is the ice skater who pirouettes faster and faster as she draws in her arms.[2]

The work of Schauberger and other maverick inventors is now beginning to be recognised by some members of the scientific community. Numerous well established Ph.D. scientists have now left their jobs in academic and research organisations to work independently on investigating new forms of energy which are not yet recognised by the scientific establishment.[3] Many of these forms have in common a link with the vortex as a connection to cleaner more sustainable energy. Some call it "free energy" and have successfully demonstrated machines that give out more energy than they take in. Some machines are even claimed to overcome gravity.

There is as yet little agreement among scientists on the name of this form of energy although some agreement is appearing on its properties. Long recognised in the East as chi or prana, this energy is now seen as being non-physical in the normal sense in that it cannot be measured by normal physical means and does not fit into the current physical model of electromagnetic energy. It is sometimes called "scalar energy" but has also been christened "zero point energy" because it is known to be still present even when the temperature of an object has been reduced to absolute zero (minus 273°C.), at which point all of its molecular activity should have ceased.

This energy state was previously called the "vacuum state" which is accessed by going to smaller and smaller levels. Nothing was supposed to exist beyond this point which was at the border of existence, at a scale of 10^{-33}cm., called the Planck scale. Anything smaller than this was considered to be beyond space and time, beyond physical existence. However, it has now been shown that this vacuum state is highly energetic and is not a vacuum at all. It is seen by many scientists as an enormous potential source of free energy. William Tiller, Professor Emeritus of Stanford University gives an indication of the amount of energy in the so-called vacuum:

> Most of the general public hold the idea that the vacuum is not only the absence of physical matter but is also devoid of anything!   However, for quantum mechanics and relativity theory to be internally self-consistent, the vacuum is required to contain an amazingly large inherent

energy density (~1094 gms/cc). This vacuum energy density is so large that the intrinsic total energy contained within the volume of a single hydrogen atom (~10^{-24} cm^3) is more than one trillion times larger than that contained in all the physical mass (mean mass density ~5×10^{-28} gms/cm^3) of all the planets plus all the stars in the entire known cosmos out to a radius of 20 billion light-years. This makes the energy stored in physical matter an insignificant whisper compared to that stored in the vacuum.[4]

In the old model of physics, atoms were considered to be isolated objects like tiny billiard balls. Atoms were made up of a nucleus with electrons orbiting around it like planets orbiting the Sun. Matter was composed of atoms and particles which were separate from an objective observer carrying out an experiment. It is now understood that everything is connected and that you cannot change one thing without affecting everything else. Matter is now seen by some scientists as a form of resistance to movement through a field of zero point energy, a mere concentration of energy rather than something physically separate.[5]

At this point it is worth looking at the work of the world famous physicist David Bohm. Bohm, who had a considerable reputation in academic circles and came to the attention of the general public in 1980 with the publication of his book *Wholeness and the Implicate Order*.[6] This proposed that behind the supposedly 'real' world that we see around us, which he called "the explicate order", lies an unmanifest world called "the implicate order" which is multi-dimensional. The things we perceive are the explicate order, the three-dimensional world of objects in space and time. These objects are made of dense matter which, although it can be described by physics, cannot be adequately explained. Most of physics operates on this basis, presenting ideas in mathematical terms which are not easily understood and are subject to different interpretations. The implicate order provides a deeper level of meaning. What is going on in the implicate order is mostly enfolded or unmanifest. What we see is only a small part of the whole.

The implicate order is a theory which deals with the whole, like the concept of a unified field of all the forces in nature, but it also says that the connections to the whole have nothing to do with locality in space and time. Instead they are explained by the quality of enfoldment. This distinguishes between that which is manifest and that which is unmanifest. There is a constant movement of unfolding or manifesting and enfolding or becoming unmanifest again.

The implicate order does not look at particles in space and time as do Newton's and Einstein's field theories but instead considers the background or vacuum state which Bohm called the "holomovement" as it is too holistic to be considered separate and too lively to be described as a vacuum. All that is manifest is basically floating in this state. Matter can be regarded as a cloud of more dense concentration within the holomovement. However, this state has been ignored by conventional physics because it cannot be detected by conventional instruments, since it is beyond space and time. It is generally agreed though, that it is a source of infinite energy. This energy is implicate, folded-up or spiralled into itself.

A field can be considered to be the area of influence of a force. Forces decay with distance so the concept of a field has traditionally included the idea that the farther away you are, the less you can influence something. However, the zero point field has the property of non-locality. It is independent of distance. Some scientists who have approached this subject from a different angle have called the zero point field by a different name. They call it a torsion field and show that it manifests in the form of vortices or spirals of energy, based on the concept of spin.

We think of spin as a property of matter. We can imagine a flywheel spinning and we know that it possesses angular or rotational momentum because it is very difficult to stop. But spin is also a property of elementary particles. Spin may be a fundamental property of nature, from the smallest to the largest objects; for example fields with spin properties are seen in the rotation of stars and galaxies. These fields are closely related to magnetic fields and we are all affected by them.

Scientists often work on parallel tracks without being aware of each other. Russian scientists have been carrying out research on subtle energy fields and psychic phenomena since the 1940s. It appears that they have not been as reluctant as their Western counterparts to embrace unusual concepts, possibly because the Soviet Government which funded their efforts was looking for military applications. Reports of this work have recently come to light in the West with the translation and publication of details of research on torsion fields which are intimately connected with the property of spin.

The research on torsion fields is reviewed in an interesting paper by Yuri Nachalov.[7] He reports that a Soviet astrophysicist named Kosyrev demonstrated in the 1940s that spinning gyroscopes can be shown to lose weight. He points out that this has been observed by a number of scientists in different countries. It was suggested that this phenomenon was caused by an unknown energy effect seen in freely spinning objects including stars and planets. This effect appeared to be related to the direction of spin and was observed more easily in objects which were themselves moving as well as spinning. This led to measurements of electromagnetic energy received from space in the form of spinning moving particles such as photons.

But when Kosyrev measured the energy received from stars he found an unusual effect. He showed that if the light or electromagnetic waves were blocked by a metal screen, images of stars were still received. Not only that, but the image received suggested that the distant star was in a different place, leading to the conclusion that the type of energy received had travelled faster than light. It also suggested that electromagnetic radiation, as we know it, is accompanied by another component, a spin or torsion (twisting) component.

A number of Soviet researchers made generators composed of spinning magnets which were claimed to demonstrate anti-gravity effects. These generators produced a spin field which was claimed to change the inner structure of any substance. Nachalov states that:

> The property which is open to influence by torsion fields is spin. (We should note that the spin-torsion interaction constant is equal to 10^{-5}- 10^{-6}. This constant is less than the constant of electromagnetic interactions, yet much greater than the constant of gravitational interactions.) Thus the structure of the torsion field of every object can be changed by the influence of an external torsion field.

He then goes on to make the very unusual statement that the effects produced were similar to effects produced by psychics, i.e. that spin fields could be used to replicate psychic phenomena and that psychics could produce changes similar to that produced by torsion fields!

It seems that there is some strong but not yet fully understood connection between spin fields and gravity. There would also appear to be some connection between gravity and psychic effects such as psychokinesis (moving objects by mental power), extra-sensory perception and healing. We will be developing this point in some detail later.

There have been a number of cases reported of heavy objects being moved by apparently miraculous means often by only one man. John Ernst Worrell Keely developed a theory of vibratory physics which held that all atomic particles were composed of vortices, a concept similar to Lord Kelvin's theory. Keely is reported to have been able to levitate large objects using devices attached to a belt around his waist. Unfortunately the reports of these feats are difficult to verify.

There is better evidence to support the work of Edward Leedskalnin, a Latvian who emigrated to the United States and built a park of stone monuments called Coral Castle which is to this day a major tourist attraction at the town of Homestead, Florida.[8] It can be seen in Figure 1 below. Like Keely who died in 1898, Leedskalnin who died in 1951 took his secrets with him to the grave.

Figure 1.
Coral Castle constructed from huge stones of up to 30 tonnes weight moved by one man, Edward Leedskalnin. How he did it is a mystery

(photo from Wikipedia commons /Christina Rutz)

38

Leedskalnin, who was only five feet tall, apparently formed and moved into position a 22 ton obelisk and other large stone blocks weighing up 30 tons and many smaller items. He did all of this on his own but was always very careful to make sure nobody was looking when he did it. Leedskalnin's workshop contained a device composed of rotating magnets which may have been an anti-gravity generator.[9] He believed that magnetism flowed in currents of tiny spinning particles which could penetrate any substance. A website devoted to Keely and other anti-gravity researchers describes a rotating magnet generator and gives the results of experiments which claim weight reductions of 25-50%.[10]

Support for these theories of torsion fields and unknown gravity effects has come from a more reputable source. Professor Eric Laithwaite formerly of Imperial College London and inventor of the linear electric motor, gave a lecture to the Royal Institution in London where he demonstrated a rotating gyroscope consisting of a motorcycle wheel connected to a shaft and rotated by an electric motor. Laithwaite showed that the apparatus weighed over 50 pounds and was very difficult to lift. Nevertheless, after switching on the motor and getting the unit to rotate at high speed, he was able to lift it above his head with one hand.

That meeting in 1973 is regarded by many as one of the most shameful events in the history of science. Laithwaite's demonstration was greeted with frosty silence since he appeared to be taking issue with the great Isaac Newton. The society failed to publish a report of the meeting and acted as if the demonstration had never taken place. Laithwaite subsequently applied for a patent for an improvement of his device, which explains how it operates. In it he states that "matter is moved without reaction and a propulsion device is established".[11] Newton stated that action and reaction are equal and opposite. You cannot push a thing in one direction without something being pushed in the opposite direction. For example if you stand in a boat and push against the bank, the boat will move outwards. Laithwaite's patent shows that his device will move upwards without creating any downward force. (See overleaf).

PCT   WORLD INTELLECTUAL PROPERTY ORGANIZATION
International Bureau

INTERNATIONAL APPLICATION PUBLISHED UNDER THE PATENT COOPERATION TREATY (PCT)

(51) International Patent Classification 4 :		(11) International Publication Number:	WO 86/ 05852
F16H 33/20	A1	(43) International Publication Date:	9 October 1986 (09.10.86)

(21) International Application Number: PCT/GB86/00172	JP, LU (European patent), NL (European patent), NO, SE (European patent), SU, US.
(22) International Filing Date: 25 March 1986 (25.03.86)	
(31) Priority Application Number: 8507684	**Published**
(32) Priority Date: 25 March 1985 (25.03.85)	*With international search report.* *With amended claims.*
(33) Priority Country: GB	
(71)(72) Applicant and Inventor: LAITHWAITE, Eric, Roberts [GB/GB]; 9 Shorecrofts Aldwick, Bognor Regis, West Sussex PO21 4AS (GB).	
(74) Agents: ROGERS, Jack et al.; F. J. Cleveland & Company, 40-43 Chancery Lane, London WC2A 1JQ (GB).	
(81) Designated States: AT (European patent), AU, BE (European patent), BR, CH (European patent), DE (European patent), DK, FI, FR (European patent), GB (European patent), IT (European patent),	

(54) Title: IMPROVEMENTS IN OR RELATING TO A PROPULSION DEVICE

(57) Abstract

A thrust producing device comprises a support (10) to which torque can be applied. A cross member (13) fixed to the support carries rotors (17) on axels (16), the axles being pivoted to the cross member. When the torque is applied with the rotors spinning a thrust is developed.

Figure 2. Professor Laithwaite's patent application for an anti-gravity device in opposition to Newton's Third Law of Motion.

Torsion fields have properties that are related to electromagnetic fields. When electric or magnetic fields are created, torsion fields are also present. Usually their effects have been masked by the stronger electromagnetic fields and so they have gone unnoticed. Where electromagnetic fields are screened, the torsion fields can be detected. Kosyrev detected torsion waves by blocking the light entering his telescope with a metal screen.

Electromagnetic waves are also screened by layers of earth and rock. This is why radios do not work underground. It is likely therefore that the 200,000 tons of rock and earth in the mound at Newgrange served to screen the cruciform chamber from all electromagnetic fields. Anything inside would be acted on by torsion fields thereby giving effect to the possibility of the structure of the torsion field of an object being changed by the influence of the external torsion field as referred to by Nachalov above who goes on to state:

> the new configuration of the torsion field will be fixed as a metastable state (as a polarized state) and will remain intact even after the source of the external torsion field is moved to another area of space. Thus torsion fields of certain spatial configuration can be "recorded" on any physical or biological object.[12]

In other words Nachalov is saying that by placing an object in an external torsion field, its own torsion field is permanently changed. Note that he says any physical or "biological" object. This means that a person spending time inside a mound like Newgrange would experience some kind of permanent change.

The crossing of underground streams under the mound may be the cause of the torsion fields or an effect of them. Since torsion fields are also related to magnetic fields, they may be closely associated with changes in the Earth's magnetism or with the phenomenon known as leylines, mysterious lines of subtle energy. Newgrange and most sacred sites are widely believed to be focal points or nodes where networks of leylines meet. Perhaps the concept of leylines, as yet not accepted by science, may come to be understood as the effect of torsion fields.

Unfortunately much of what Nachalov says is difficult to verify and very few of his ideas have been taken up by Western scientists. Indeed considerable scepticism has been expressed on the subject of torsion fields by the Presidium of the Russian Academy of Sciences. However, this refers to the claims of two Russian scientists A. E. Akimov and G. I. Shipov and does not refer to the earlier work described above by Nachalov. Many of the papers Nachalov refers to are still only available in Russian but some other Western scientists appear to be coming independently to similar conclusions as we shall see in the next chapter.

Chapter Five

Overcoming Gravity

Professor Laithwaite, as we saw in the last chapter, has shown how spinning devices can produce anti-gravity effects. This illustrates a fact little known outside the physics community. That fact is that we do not understand gravity. We know about the so-called 'Law of Gravity' because we observe that objects fall towards the earth, but even scientists do not understand the nature of this mysterious force. Newton said that inanimate objects could not possibly attract each other and concluded that gravity must be a force of divine origin. Einstein told us that gravity was caused by the curvature of space, but this does not explain the actual mechanism of attraction and Einstein's approach is now being subjected to detailed scrutiny and revision.

One of the problems in understanding gravity is that it is really a very weak force. Compared to the electromagnetic force it is only 0.23×10^{-42} times as strong. That's 0.23 but with 42 zeros after the decimal point, so small as to be difficult to measure. However, we know how to calculate the gravitational interaction mathematically. We know that the force of attraction between two objects is proportional to the mass of the objects and decays as the square of the distance between them. Gravity only appears large to us because we are on the earth which is so massive when compared to us. Between objects of similar size on earth, the force is negligible.[1] Maybe gravity is not that difficult to overcome, and once overcome, the laws of physics tell us that no energy is required to keep an object suspended in mid-air.

We tend to think of the enormous energy required for aeroplanes to stay in the air and we see that even helicopters have to keep their engines running to hover. But these are inefficient methods of flight. Hovering is done by using energy to fly upwards against gravity which is drawing the helicopter downwards. Gravity is a force not an energy, so if we can find an equal and opposite force we can counteract gravity without using energy. A helium balloon stays suspended in mid-air with no energy input because it is lighter than the air it displaces and the buoyancy of the air counteracts gravity.

Many people have seen the demonstration of overcoming gravity using subtle energy in the hands where a person is lifted off a chair. A person is put sitting in a chair and four volunteers each join their hands with their index fingers protruding and their other fingers folded over. They are then asked to lift the person by putting their fingers, still held as before, under the person's knee joints and armpits. They are unable to do it. They are then asked to place their open hands on the person's head. All the hands are now on top of each other with alternate left and right hands in a vertical pile on the person's head. Next they are asked to quickly join their hands together as before and try to lift the person. This time the person easily rises into the air for a few seconds until the energy field decays.

I have done this demonstration with groups of people and it has always worked. On one occasion we had eight people, male and female and not especially muscular, sitting around a heavy boardroom table. First I asked them to put their hands under the table top and try to lift it off its base. With some struggling some of the men lifted it a few inches. Then I asked the group to put their hands palms down on the table for a few minutes. I then told them to quickly lift the table again and they did it with ease. I have read that this technique has been used to get a car out of a ditch!

The fact that we do not fully understand gravity is further illustrated by our inability to duplicate the feats of construction of the Neolithic monument builders. We do not know how the heavy stones were moved. Even with today's technology we would have great difficulty in matching some of their their achievements. To get a deeper understanding of gravity and its relevance to Neolithic construction we should look first at what the Einstein revisionists are saying.

Albert Einstein published his General Theory of Relativity in 1915. This concerned itself with gravity and the curvature of space-time. His earlier Special Theory of Relativity proposed in 1905 was based on his ideas about the speed of light. In the 1920s the first ideas of Quantum Theory emerged and scientists pursued this area to the detriment of further work on General Relativity. In fact although quantum mechanics is consistent with Special Relativity, it has never been reconciled with General

Relativity. In the so-called 'Standard Model' of physics, gravity remains on the outside, the poor relation, a weak, little understood force which resists attempts by physicists to fit it into the Holy Grail of physics, a Unified Field Theory or a theory which unites all the forces of nature.

In 1864 James Clark Maxwell showed that electricity and magnetism are really aspects of the same force and thus succeeded in uniting the forces of electricity and magnetism to produce electromagnetism, a force which with its accompanying mathematical equations has become the basis of radio transmission and has helped us to understand the properties of light which is an electromagnetic wave like a radio wave, but at a much higher frequency. The other forces are the weak and strong nuclear forces and gravity. Substantial work on the unification of electromagnetism and the weak nuclear force was completed in 1967 by the physicists Glashow, Salam and Weinberg resulting in the electro-weak force. More recent work has concentrated on the so-called "grand unified theories", attempting to unify the electro-weak force with the strong force. The next stage is to propose a solution which includes gravity. So far this has not been achieved and this approach seems to be running out of steam.

A different approach has been taken by Professor Myron Evans, another scientist who, like Laithwaite, has paid the price for deviating from the establishment. Evans was a professor in the University of Aberystwyth in Wales. In 1983 his research group was disbanded even though he had earlier been awarded the Royal Society of Chemistry's Harrison Memorial Prize and Meldoal Medal (among the most prestigious awards in British chemistry). He has since set up his own Alpha Institute for Advanced Study (AIAS) [2] where he pursues his researches on a meagre budget.

Evans, a chemical physicist, has proposed that electromagnetism is the torsion of space-time. His theories are based on the earlier work of the French mathematician Elie Cartan who exchanged ideas with Einstein. Cartan worked out that electromagnetism could be derived from the geometry of space-time. Einstein had derived gravity from the geometry of space-time. Thus a connection between gravity and electromagnetism was suggested which was built on by Evans. He has developed his Covariant Unified Field Theory by discovering a new torsion field called the B(3) field. A description of this is beyond the scope of this book but can be reviewed in the references provided by AIAS.

The implications of Evans' work are that energy can be extracted from the enormous reservoir of the vacuum and that this can solve a number of problems such as power generation and the overcoming of gravity. The obtaining of energy from the vacuum is stated to be a resonance effect.[3] This appears to confirm the work of Leedskalnin who appeared to move heavy objects using an electric generator which he may have tuned to the appropriate resonant frequency.

The phenomenon of resonance is one which is well known and its power is considerable. In my early career as an electrical engineer I visited a power station where a thirty megawatt turbo-alternator had developed a resonance caused by a fault in the alternator. It had simply exploded and bits had flown in all directions punching holes in the roof and walls of the power station. It is well known that marching soldiers are told to break step when crossing a bridge. This is required because the power of waves or forces which are in phase with each other is much greater than that produced by those that are randomly occurring.

Christopher Dunn, an engineer and author of *The Giza Power Plant,* suggests that the Great Pyramid was a giant energy device driven by resonating acoustic waves and states that its constructors must have used power driven machine tools, otherwise it would not have been possible to produce either the quantity of stone or the precision made surfaces found inside the King's and Queen's Chambers. Describing the Great Pyramid as "the largest and most accurately constructed building in the world", Dunn estimates that, using today's technology, it would take twenty seven years just to quarry and deliver the stone. He says that thirty Empire State Buildings could be built with the estimated 2.3 million blocks.[4]

There is still no realistic explanation as to how the stones were placed in position although many archaeologists and engineers have studied this amazing achievement. One estimate is that 1,800 men would have been needed to haul each block up ramps with a gradient of 1 in 10. These ramps would have been 4,800 feet long, three times more massive than the pyramid itself.[5] Imagine how long it would take to haul 2.3 million of these blocks, some of them weighing 200 tons.

It has been suggested that the blocks were floated to their positions on barges, but this raises the question of the enormous canals and dams that would have been required and is clearly impractical. Another suggestion is that the blocks were made of a type of artificial stone and cast in place but this does not explain the tool marks on many of the blocks. It is possible that some other form of energy was used and that this energy had been developed earlier in the building of megalithic monuments in Ireland and elsewhere. For example the *Grand Menhir Brisé* ("Great Broken Menhir") at Locmariaquer, Brittany, has been estimated to weigh 330 tonnes. It is now lying on its side broken into four pieces but it once stood 20 metres high and is thought to be the heaviest object ever moved by humans without powered machinery.

An interesting account of overcoming gravity is contained in a book called *The Bridge to Infinity* by Bruce Cathie.[6] In this case it is clear that some unknown source of energy must be involved. Cathie reports an account of levitation of blocks of stone by Tibetan priests using trumpets and drums. The account is claimed to be a translation from a German magazine article in which a Swedish doctor named Dr. Jarl was asked by a Tibetan friend to go to Tibet to treat a high Lama. While there he saw monks building a rock wall in front of a cave in a cliff some 250 metres above the ground. He claimed that blocks measuring 1.5 metres long by 1 metre wide were raised off the ground and slowly transported to the cave entrance by monks making a ferocious din with 13 drums and 5 trumpets.

The levitation of stones by Sufis in India is reported by V.S.Gopalakrishnan.[7] He shows photographs of eleven Sufi men raising an 80kg stone off the ground by placing their index fingers on it and chanting. There have been many reports of Indian holy men who could rise up off the ground by mental power alone. This is not as strange as it sounds because these techniques are now taught to advanced meditators and I have successfully practised them and can vouch for the fact that in a deep state of meditation you can lift your body off the ground for a few seconds. This has been demonstrated as "yogic flying" and has been dismissed as using muscular effort. No effort of any kind is involved. The technique is described in the Yoga Sutras of Patanjali.[8]

I can understand that many people would be sceptical of some the above accounts and of the work of Professor Evans. However, it is clear that many academics are taking Evans' work seriously and he claims to have had contact with many universities and commercial organisations who want to look into the implications of his work for the development of new technologies for transport and energy. He also claims that his theory has explained a two previously puzzling effects, the Aharonov Bohm Effect discovered by David Bohm and his collaborator Yakir Aharonov and the Inverse Faraday Effect.

The Aharonov Bohm effect showed that an electromagnet could produce an effect in regions where its magnetic field was previously considered to be absent. The Inverse Faraday Effect showed that matter could be magnetised by a circularly polarised beam of electromagnetic radiation. This could be similar to the effect described by Nachalov in the last chapter where he stated that the structure of the torsion field of any object can be changed by the influence of an external torsion field.

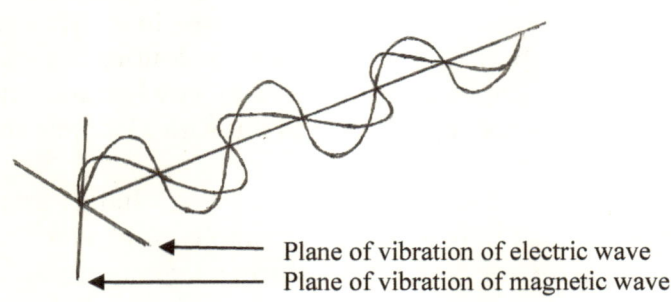

Plane of vibration of electric wave
Plane of vibration of magnetic wave

Figure 1. An electromagnetic wave: when the planes rotate this produces circular polarisation.

An electromagnetic wave consists of a wave of electricity vibrating or resonating in one direction and a similar wave of magnetism vibrating in a plane at right angles to the electric wave. Looking at Figure 1 opposite, it can be seen that the electric wave is vibrating in a horizontal plane while the magnetic wave is vibrating in a vertical plane. Now imagine that an extra component is added to the waves by causing the planes to rotate at the same time as the waves are propagating. This is called circular polarisation. The effect produced is one of torsion or twisting.

We have seen that there are energy effects which are not yet fully understood. An extra component of electromagnetic waves is now being studied and some scientists such as Evans seem to have an explanation which although not mainstream is attracting significant scientific interest. This extra component is a torsion field and is related to gravitational effects which have not been previously explained. By the use of resonance it appears that it could be possible to levitate objects by tapping into the free energy of the vacuum. This could explain how enormous stones were moved to build the Neolithic monuments. The spiral designs which are evident in the later Irish monuments could be a representation of this form of energy and it could have had a number of uses. In later chapters we will explore these ideas further and try to understand this energy and its applications.

Chapter Six

Ancient Connections

I have described the two models of society – the dominator model which we have at present and the partnership model which is believed to have existed in Neolithic times. The dominator model is reductionist and based on the principle of 'divide and conquer'. The partnership model is holistic. A dominator society is secretive. It is afraid that others will gain access to its knowledge and use it in a hostile manner. A partnership society is likely to share its knowledge since it does not fear attack. Recent archaeological discoveries have led to theories which suggest that there was widespread communication between ancient societies and that early civilisations were more than just isolated pockets of higher consciousness as previously supposed. This leads to the conclusion that the principle of partnership was widespread and supports the theory that partnership societies were likely to have been the cradle of civilisation.

Part of our evolution as a species may consist of finding out the truth about our origins and benefiting from this ancient knowledge. There have been so many new archaeological discoveries in recent times that it is possible that these may be linked to the general increase in spiritual awareness which is taking place outside a religious context. In this chapter I will attempt to give an overview of the recent discoveries and associated theories. Some of these theories are highly controversial and are the subject of ongoing debate between scholars. Others are coming from outside the academic tradition and are being treated with great scepticism by the establishment. In considering these theories I have tried to keep an open mind. However, this does not mean that we should greet every new theory as being proven fact. Where possible I have tried to look for consistency and support between the various ideas and have rejected those many others which seem to be only standing on one leg.

There is now general agreement that when the last Ice Age ended around 11,000 B.C. there was a considerable rise in temperatures and in sea levels around the world as the ice melted. Many areas disappeared beneath the sea. Working backwards it has been possible to see how countries presently separate were once connected by land bridges. A good example is Alaska and Russia. These land bridges facilitated the migration of people and animals and suggest that the Americas were populated by migrants from Asia. However there are many discoveries that cannot be explained by such theories. For a long time archaeologists have been puzzled about finds of remains of tropical plants and animals in polar regions. For example, coal and fossilised tree stumps have been found in Antarctica and fossilised palm leaves have been found in Spitzbergen, an island north of Norway. Fossils go back much further than 11,000 B.C. but the finding of them shows that the polar and tropical regions of the earth have not always been in the same places as they are now.

Fully preserved deep frozen remains of mammoths and other large animals of more recent origin have been discovered in Siberia. These animals would have required extensive vegetation to survive, thereby suggesting that Siberia was once a temperate zone. The origins of many of the cultivated seeds used in agriculture have been traced back to mountainous regions widely scattered over the earth. One would expect their origins to be concentrated in a small number of places where there are fertile plains. What is the explanation for these apparent anomalies? Rand and Rose Flem-Ath offer an explanation in their well researched book, *When the Sky Fell: In search of Atlantis*.[1] Their explanation is based on the theory of "earth crust displacement".

There have been numerous theories proposed to explain the apparent shift of climatic regions of the earth. One theory now out of favour was that the earth flipped and its axis changed from north-south to an east-west position. This is based on gyroscopic action and can be understood by comparison with a spinning top. When it is first spun, the top spins on a north-south axis, but then it begins to wobble and eventually flips onto its side and continues to spin on an east-west axis. This theory required the whole earth to wobble and flip and it is unlikely that any life could have survived such a catastrophic event. The later theory of earth crust displacement is much more plausible and had the support of no less a scientist than Albert Einstein. According to this theory the earth's crust slides on the molten magma layer about eight miles underneath it, as explained by Einstein in a foreword to a book called the *Earth's Shifting Crust* by Charles Hapgood.

In a polar region there is a continual deposition of ice, which is not symmetrically distributed about the pole. The earth's rotation acts on these unsymmetrically deposited masses, and produces centrifugal momentum that is transmitted to the rigid crust of the earth. The constantly increasing centrifugal momentum produced in this way will, when it has reached a certain point, produce a movement of the earth's crust over the rest of the earth's body, and this will displace the polar regions toward the equator.[2]

The shift of the earth's crust should be thought of as like moving a completed jigsaw puzzle. You can slide the whole puzzle across the table if you push at one edge. Because the friction between the puzzle and the table is low it will not buckle or come apart. Similarly the earth's crust could slide in relation to its core thus relieving the imbalance caused by the build-up of ice at the poles. This does not mean that the shift of the earth's crust occurred gently. It would have been accompanied by earthquakes and enormous tidal waves which would have swamped all low lying land masses. People living in mountainous regions would have stood a better chance of survival. It is now believed by many scientists that this phenomenon has occurred regularly in the past, brought about by a combination of three astronomical cycles. These are the tilt of the earth's axis, the elongation of the earth's orbit and the closeness of the earth to the sun in summer. These cycles coincide every 41,000 years, with the last displacement likely to have occurred in 9,600 B.C. shortly after the end of the last Ice Age.

The last shift of the earth's crust is thought to have moved the poles by about 30 degrees. Before the shift the North Pole is believed to have been located in Northern Canada where Hudson Bay is now. The movement of the ice cap would have caused some of it to melt, thus raising the level of the seas, and this may be an explanation for the flood mentioned in the Bible. The seas would only have receded slowly as the new ice cap formed. Not all areas would have been equally affected. The movement would have taken place about an axis passing through Africa and the Pacific and these regions would have been the least affected. Siberia, in contrast, would have moved north by about 30 degrees. In other words it would previously have been 30 degrees further south, close to the position presently occupied by central China today and would have experienced a temperate climate. The movement would have taken place in a short period of perhaps a few days. For the animals in Siberia, the sudden climate change would have trapped them in a deep freeze. When the seas swamped the plains, agriculture would have been wiped out and this explains how the origins of cultivated seeds have been traced to widely dispersed mountain regions.

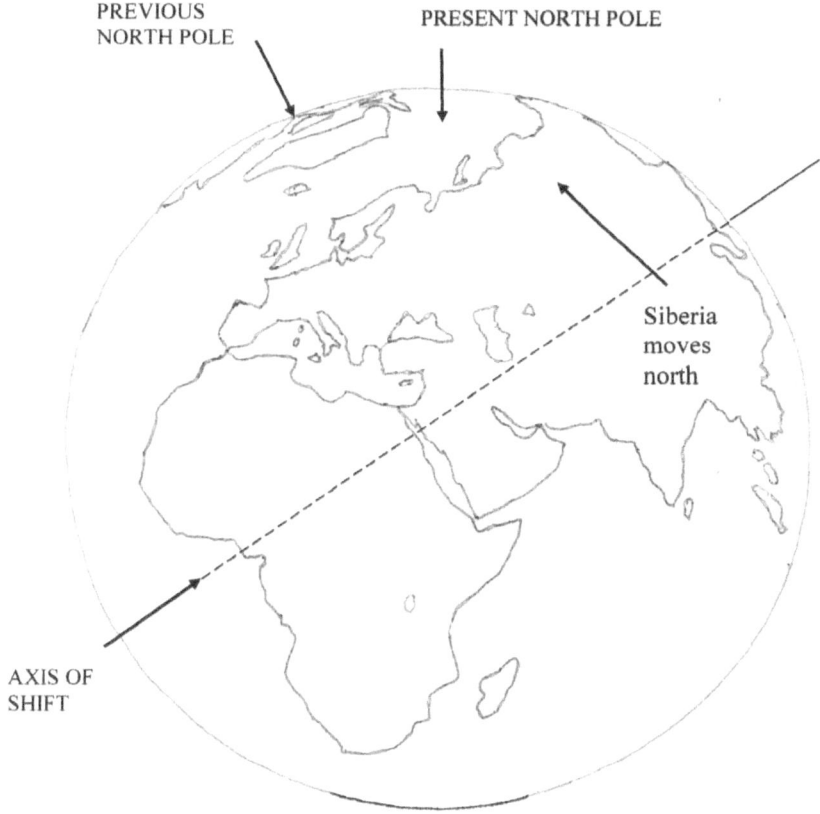

Figure 1. The earth crust displacement would have taken place about the axis shown thereby causing Siberia to move from a temperate latitude in relation to the previous pole to a more northerly one in relation to the present pole.

The theory also explains why there are huge ice fields in regions of Greenland and Antarctica although little snow falls there today. These regions would have been close enough to the poles to have acquired their ice fields before the shift. The part of Antarctica closest to South America would have been located about where Southern Chile and Argentina are today and this is the location proposed by Rand and Rose Flem-Ath for the lost continent of Atlantis. According to them it did sink under the waves but when the seas receded it remained covered in ice.

There have been over one thousand books written on the subject of Atlantis, most of them proposing new locations for the vanished continent, and I was surprised to find when I started researching this subject that many academics, especially historians and archaeologists, have devoted considerable time to it. It has become almost respectable as a research area since the discovery by Professor Spyridon Marinatos of the Minoan settlement on the Aegean island of Santorini, which I referred to in Chapter 2, and which was annihilated by a volcanic eruption in or about 1629 B.C.[3] It was suggested that what had existed prior to the eruption bore a resemblance to the detailed account of Atlantis given by the Greek philosopher Plato who lived from 429-347 B.C.

Riane Eisler in her book *The Chalice and the Blade* quotes a number of references in support of the theory that the Minoan civilisation was Atlantis. Quoting from *The End of Atlantis* by J. V. Luce, she says that some of the elements of Plato's Atlantis are a "startlingly accurate sketch of the Minoan empire at the sixteenth century B.C.E." (1,700 B.C.) She goes on to say that it "seems that the story of Atlantis is actually the garbled folk memory, not of a lost Atlantic continent, but of the Minoan civilisation of Crete".[4]

Considerable research was undertaken by Andrew Collins for his book, *Gateway to Atlantis* [5] and he quotes extensively from Plato's works. In one of his books called the *Timaeus* Plato relates how Solon (c. 638-558 B.C.), a celebrated Athenian legislator, had travelled to Egypt and heard of an ancient civilisation from an elderly Egyptian priest who stated that there had been many floods in the past.

Plato's *Timaeus* seems to suggest that the Egyptian civilisation then was at least 8,000 years old and that its records went back that far. This ties in with some recent arguments that the Sphinx is actually much older than was previously understood. This is based on studies of the weathering of the Sphinx, which suggest that the uneven weathering was produced by rain and not by sand.[6] It is now accepted by academics that there was a different climate in Egypt in the distant past. It has been suggested that the Sphinx was originally a lion but that the later (dominator?) society of pharaohs had the lion's head recarved in the image of a pharaoh.[7] It can be seen by comparing the two photographs below that the pharaoh's head is much smaller than the head of a lion and that this may have been a consequence of the modifications. In converting the lion's head to a pharaoh, they had to carve away so much that the resulting pharaoh's head was smaller than intended.

Figure 2. The dominator society of pharaohs in Egypt may have carved the head of a lion to make the Sphinx in the image of a pharaoh. There is ample evidence that the Sphinx is much older than the pyramids and dates from a time of temperate and rainy climate thousands of years before the pharaohs.

(photos from Wikipedia Commons/Rupal Vaidya and Usuario:Barcex)

Plato also described bull sacrifice as one of the rituals of the Atlanteans and this attracted a lot of interest from academics who saw parallels with Minoan Crete. When the Minoan palace at Knossos was discovered by Sir Arthur Evans in 1895, "he discovered walls adorned with enormous frescoes that depicted in vivid detail young men and women performing the dangerous act of bull-leaping, while elsewhere bull motifs and symbols were found in profusion".[8]

It is beyond the scope of this book to delve any deeper into the Atlantis mystery but it is interesting to note that Crete is on a major fault line and suffered from earlier earthquakes. Thus even if Atlantis was not destroyed by the movement of the earth's crust, there are other theories to explain how it may have declined. It may have been destroyed by earlier earthquakes or by rising sea levels. At the end of the Ice Age the sea level was some 400 feet lower than it is now, and the Sea of Crete is dotted with islands, many of which would have been joined together in the past when the sea levels were lower. It is therefore possible that Atlantis may have existed in the Mediterranean, the 'known world' of Plato's time, but that is that it suffered a gradual decline as the seas rose after the Ice Age. It was destroyed either by earthquakes, rising seas or the crust displacement of 9,600 B.C. leaving as a vestige, the Minoan civilisation, which in turn was destroyed by the Santorini eruption of 1629 B.C. referred to in Chapter Two

The title of the book by Rand and Rose Flem-Ath, *When the Sky Fell* is a reference to the movement of the sky during the crust displacement as it must have been seen by those who witnessed it and survived. They relate the well known story that the Celts had no fears except that 'the sky might fall on their heads'. It is amusing to note that this has since been popularised in the person of chief Vitalstatistix who appears in the children's books about Asterix the Gaul. The Flem-Aths go on to suggest that Neolithic monuments may have been astronomically aligned to provide a method of measuring the sun's position and reassuring the people that the sun was following its correct course and that the sky was not likely to fall on their heads again.

At Carnac in Brittany there are parallel rows of standing stones which stretch for a distance of over two miles and are aligned with the path of the sun in summer. There are over two thousand stones arranged in twelve rows running approximately east-north-east to west-south-west. Many of the stones are over fifty tons in weight.[9] Perhaps these stones were part of some ritual or energetic guiding of the sun to remain on the correct path. One cannot help being impressed by the fact that throughout Ireland, Scotland and the western parts of England and France, millions of tons of stones were moved to create vast monuments with some astronomical purpose. Stonehenge was built in a number of stages, the last one being finished in about 1,600 B.C. This would appear to be the last construction of a highly organised society preoccupied with astronomical alignments. This preoccupation is no longer seen after the coming of the Bronze Age which brought in the technology for making swords.

The Flem-Aths also relate an interesting fact about Bal Gangadhar Tilak, an Indian Sanskrit scholar who in 1903, published his most famous work *The Arctic Home in the Vedas*. In this book he argued that an ancient Vedic paradise existed in the Arctic and that the ancient texts described a flood and a story similar to that of Noah's Ark.

If there was a displacement as described, perhaps it explains the origin of the Vedic civilisation. After the shift the survivors would have moved south and may have settled in the Indus Valley. Perhaps these were the blue-eyed and fair skinned Aryans whose mysterious appearance in this region has puzzled historians. The discovery in recent times of the ruins near Harappa in Pakistan has revealed an ancient civilisation on a par with that in Minoan Crete. The city at Mohenjo-Daro was built on mud platforms to protect it from floods, had multi-storey houses, was laid out on a grid pattern and had sewers and toilets like a modern city. A large number of female statuettes were found, suggesting worship of a goddess. Further excavations have revealed remains of more cities and satellite photography of the Thar Desert has shown traces of a great river and canals. Perhaps this is the mythical river, the Saraswati, which is mentioned in the *Rig Veda,* and whose location or actual existence has been a subject of discussion among scholars.

The similarity between the mound at Newgrange and the Minoan Treasury of Atreus is referred to by George Coffey in his book *New Grange and other incised tumuli in Ireland.*[10] He also refers to the similarities in the spiral and lozenge ornaments found in Newgrange, in Minoan Crete and in Scandinavia. He states the view that the Minoan influence came to Ireland via Scandinavia and not via France and Britain as others have proposed. Other research shows that sea travel was more common than had been appreciated previously, since it was much easier to move goods by sea than over land which had dense forests and no roads.

In addition, similarities have been demonstrated between the Irish stone monuments and their decorations and those in France and Morocco. It is known that Moroccan sailors travelled to the Eastern Mediterranean and perhaps to Crete. The Neolithic monuments in Malta have spiral designs as well as the famous 'fat ladies' suggestive of a society that honoured the female principle of fertility. Similarities have also been observed between the level of sophistication of Minoan Crete and the ancient Vedic civilisation of India as mentioned above, although their art and writing have not so far been seen to share the same origins, even though their language, Sanskrit, does. In 1786 Sir William Jones demonstrated that Sanskrit was related to Greek, Latin, German and the Celtic languages. Many people have commented on the similarities between the Irish language (Gaelic) and Sanskrit. Both have a preponderance of "a" sounds and the word *sean-scriobh* in Irish (pronounced shanscreev) means "old writing".

In this chapter I have described how the Neolithic society which included Newgrange developed after the last Ice Age and the possible earth crust displacement which followed. As climates changed some of the surviving people migrated north and repopulated the areas of north-western Europe. They settled down and became farmers and had time to specialise in more esoteric pursuits such as astronomy. I believe that the folk memory of the day the sky fell became ritualised in the worship of the heavens and the gods of the sun and the moon. Major festivals were organised around the solstices and it is possible that the winter solstice at Newgrange was celebrated with the rising sun being seen to halt its south-easterly drift and begin to move back towards the north-east signifying that all was well for the coming year.

Chapter Seven

Layers Within Layers

We can take the theory of earth crust displacement described in the previous chapter and apply it to the work of Dr. James DeMeo. In his book *Saharasia: The 4000BCE Origins of Child Abuse, Sex-Repression, Warfare and Social Violence in the deserts of the Old World*,[1] DeMeo argues that the dominator model of behaviour had its origins in the arid and semi-arid regions stretching from the Sahara through Arabia and Iran to Afghanistan, a region which he calls "Saharasia". He has collated detailed anthropological data for violent and repressive behaviour and claims that this data, combined with archaeological research, shows that dominator behaviours developed in these regions as a result of gradual climate change. He claims that none of these behaviours existed before 4000 B.C. and that the transformation took place at the same time as a major climate change "from relatively wet to arid conditions". The increasing scarcity of food and water caused by the climate change, led to competition among the population and to the emergence of the dominator society as the survivors.

While earth crust displacement might have been the original cause of the climate change, it may have taken thousands of years to affect Saharasia which, being closer to the axis of crust shift, experienced less displacement but may have been affected by the gradual establishment of a new pattern of weather systems. In the diagram overleaf it can be seen that the earth crust moved the North Pole from Hudson Bay to where it is now. This movement swivelled about an axis in West Africa and areas close to this axis would have experienced relatively little movement. We saw in the last chapter that Egypt had a much wetter climate in the distant past as shown by the weathering of the Sphinx, so it would have taken some time for the present weather patterns to become established and for Saharasia to become as arid as it is today.

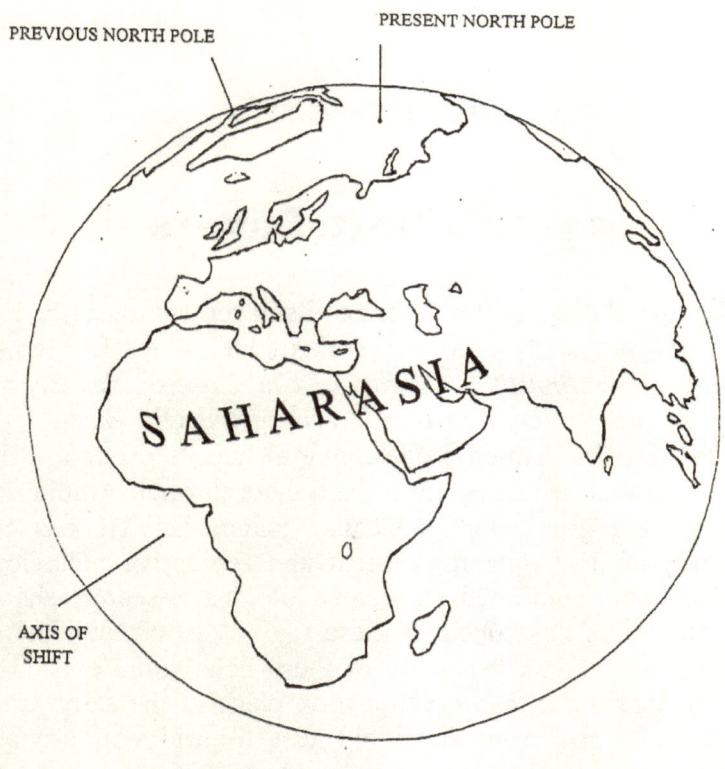

PREVIOUS NORTH POLE

PRESENT NORTH POLE

SAHARASIA

AXIS OF
SHIFT

Figure 1. Saharasia: the area where dominator behaviour possibly resulted from the pole shift during the earth crust displacement.

It is now well accepted that climate change from relatively wet to dry conditions took place in the Sahara region. Dr. Nick Brooks of the University of East Anglia has described these changes and argues that climate change was the basis for major changes in ancient civilisations. "What we tend to think of today as civilization was in large part an accidental by-product of unplanned adaptation to catastrophic climate change," says Brooks. "Civilization was a last resort - a means of organizing society and food production and distribution in the face of deteriorating environmental conditions." [2] This sounds like the beginning of the dominator society. Certainly it may have been the basis for increased fortification as people sought to protect themselves from immigrants arriving from drier areas and it is interesting to note that Saharasia, which includes North Africa, the Middle East, Iran, Iraq and Afghanistan continues to be the main source of unrest in the world, This may also be related to the type of subtle energy associated with particular places.

DeMeo's work is controversial because he based it on the theories of Dr. Wilhelm Reich. Reich was born in 1897 in Austria and died in 1957 in prison in the United States. Reich studied with Sigmund Freud but parted company with him, forming his own theory that the libido was a form of energy rather than a desire or instinct. He called this energy "orgone" and described its properties in terms which align closely with the type of torsion field or similar kinds of energy discussed in previous chapters. This energy is universal and has no mass. It is a medium for electromagnetic and gravitational energy, thus suggesting that they are specific forms of orgone.

In fact, Reich believed that static electricity was a measurable form of orgone. He claimed that this form of energy had negative entropy, thus opposing the more usual tendency of energy to run down and dissipate. This negative entropy, according to Reich, caused the concentration of orgone energy possessing organising power to produce life. Separate streams of orgone energy were claimed to superimpose on each other forming spiral concentrations in Nature and producing galaxies, solar systems and hurricanes.

After living in Austria, Germany, Denmark, Sweden and Norway, Reich eventually settled in the US state of Maine in 1944. There he founded a research organisation and produced publications and energy devices. One of these, his orgone accumulator, was said to have health benefits and his "Cloudbuster" was claimed to be able to regulate the weather. He was a subject of some suspicion to the authorities and was served with an injunction forbidding the transport of orgone accumulators across state boundaries. Because this was disobeyed by one of his staff, Reich was jailed for criminal contempt of court. His books and papers were burned by the U.S. Food and Drug Administration and he died in prison shortly before he would have been released on parole with the condition that he was forbidden to write or work on orgone again.[3]

61

Reich was in favour of natural parenting since he believed that the destructive and violent aspect of human behaviour was an abnormal condition resulting from the inhibition of natural tendencies such as emotional and sexual expression. This inhibition he believed had become a form of conditioning resulting from repressive attitudes in families due to religious and social values. These views, while embraced by liberals, were strongly opposed by conservatives and caused Reich's work on orgone energy to be treated with suspicion and hostility.

DeMeo is director of the Orgone Biophysical Research Laboratory (OBRL) in Ashland, Oregon. He has been investigating the work of Wilhelm Reich since 1970, and founded OBRL in 1978. With assistance from a number of professionals and institutes supportive of Reich's original discoveries, OBRL has grown to become one of the world's primary centres for research and educational programs focused upon Orgonomy, the science of orgone energy functions in nature.

Starting in 1977, as part of his graduate research at the University of Kansas, DeMeo undertook replication studies of Reich's biophysical research and carried out a systematic evaluation of the Reich Cloudbuster which yielded positive results. The acceptance of DeMeo's work by the faculty of the University of Kansas is considered to be the first time any aspect of Reich's controversial biophysical research had been validated by peer-review within a mainstream academic institution.

DeMeo has since carried out tests on the Cloudbuster, claiming to have successfully ended droughts across the USA and overseas. A number of desert greening expeditions have also been organized and directed by him within the arid zones of the Southwestern USA, and into the dry regions of Namibia and Israel. It is claimed that these have verified Reich's earlier findings on the ability of the Cloudbuster to bring rains under even extremely dry conditions. DeMeo claims to have undertaken a five-year desert-greening experiment in the 1990s with the support of local governments in the East African Sahel region adjacent to the Sahara Desert. All of these projects are claimed to have produced significantly positive results with increases in rainfall, ending dry episodes of long duration, filling reservoirs and greening parched landscapes.

Figure 2. An orgone accumulator
(photo reproduced with permission from Joachim Trettin, www.orgonakkumulator.de)

Reich claimed that orgone energy was attracted by organic materials and radiated by inorganic materials. His orgone accumulators, a type of box made from alternating layers of plywood and metal sheeting, have been shown to produce a rise in temperature on the inside and many have claimed health benefits from sitting inside them.[4] DeMeo has carried out tests which show that there is a spontaneous rise in temperature inside an orgone accumulator located in a controlled constant temperature environment. This cyclical temperature rise peaks at noon and would appear to mirror the torsion field component of the electromagnetic radiation received from the sun.[5]

These structures may work on the principle that energy is attracted to the organic material on the outside and then charges up the structure as if it was an electrical capacitor having alternate layers of organic and inorganic material. One could extend the capacitor analogy by saying that the organic material is the conductor or capacitor plate and the inorganic material is the insulator or dielectric.

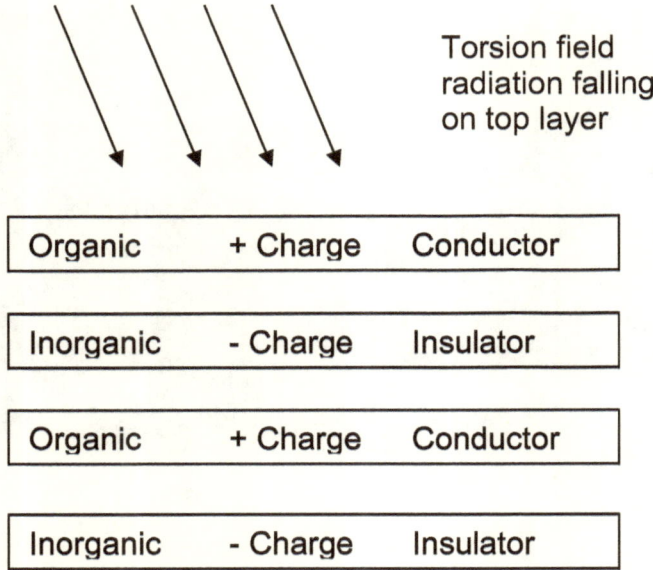

Torsion field radiation falling on top layer

Organic	+ Charge	Conductor
Inorganic	- Charge	Insulator
Organic	+ Charge	Conductor
Inorganic	- Charge	Insulator

Figure 3. Orgone energy is accumulated inside an orgone accumulator by the charging up of the successive organic layers by torsion energy, a component of electromagnetic energy received from the sun. This is similar to a battery or capacitor. Here alternate layers are shown as positive and negative

It is interesting to note that when Professor Michael J. O'Kelly of University College Cork, carried out his excavations at Newgrange he stated that 'the mound had a layered structure consisting of stones interspersed with layers of turves…Where the turf layers occurred they were separated by thick layers of loose stones with no soil in the interstices'.[6] This unusual construction of alternating organic and inorganic layers has not been explained by archaeologists. It appears that the mound at Newgrange was constructed as an orgone accumulator and was designed to have healing and energising properties [7] and that these may have enhanced its use for entering trance or meditative states.

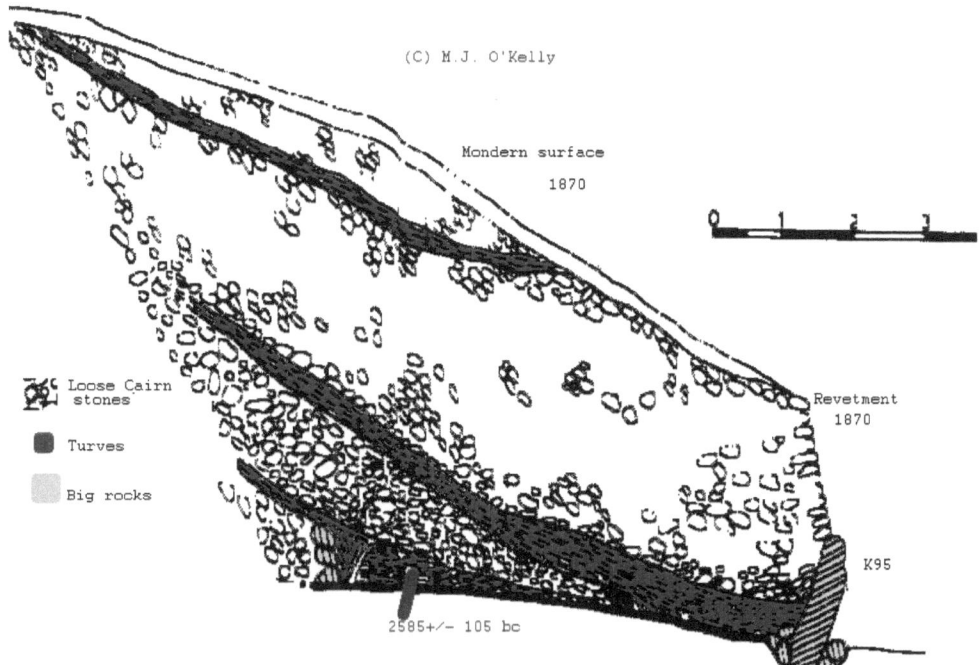

Figure 4. This drawing from Professor Michael O'Kelly's book *Newgrange Archaeology, Art and Legend,* shows a section through the mound at kerbstone K95. The alternate layers of organic (turves) and inorganic (stones) materials can be seen suggesting that the mound functioned as an orgone accumulator. The uppermost layer is insulated from the ground by the kerbstone. *(Reproduced with permission from the estate of Michael J O'Kelly)*

In making an electrical capacitor you must ensure that all the plates are well insulated from each other, especially at the edges of the plates. Usually the insulation completely surrounds the ends of the plates. If Newgrange was an orgone accumulator which worked on the same principles as an electrical capacitor you would expect that the organic layers would be insulated from each other by inorganic layers and that edges of the top organic layer, the sods of turf on the top of the mound, would be insulated from the surrounding ground. This is exactly what we find, as the kerbstones act as insulating supports to ensure that there is no short-circuiting of the upper sods to ground.

The drawing above, reproduced from Professor O'Kelly's book shows the alternating layers and the kerbstone K95 in its original tilted position. It can be seen that the topmost turf layer is insulated from the ground by K95 and the quartz wall. This wall exists only at the entrance to the mound. At the sides and back where the quartz wall is absent, the other kerbstones alone provide the insulation.

I would like to end this chapter with an observation. For many years I have felt the need to get away from the city into the country at weekends and have shared a love of walking in the country with my wife Lynda and members of a walking club. I believe that living in cities causes a build-up of stress in the individual and that this is aggravated by the lack of organic material drawing in good energy. The worst urban deprivation arises in areas of endless concrete unrelieved by parks and gardens or trees. We recognise this when we speak of "concrete jungles" and contrast them with "leafy suburbs". I was not surprised when I heard the well known herbal medicine practitioner and broadcaster Jan deVries state that he found that people who lived in houses with concrete floors had a slower recovery from illness than those who lived in houses with wooden floors.[8] It would be interesting to see some research carried out on this subject.

Chapter Eight

Stages of Development

I have stated that the dominator model described by James deMeo and Riane Eisler probably emerged due to climate change which put pressure on societies due to dwindling food and water resources as areas became more arid. This may have been caused by an earlier earth crust displacement, a concept taken seriously by Einstein. The earth crust displacement would have been witnessed by those who survived it as a massive shift in the position of the sun and the stars or as "the sky falling on their heads". This cataclysmic event would have survived as a folk memory and it is not surprising if rituals developed to appease the sun god and make sure that the sun did not stray from its path.

This would certainly explain the existence of the light box and the alignment of Newgrange and other sites where the phenomenon of the sunlight reaching a certain point and then retreating could have been an event of great significance and celebration. We tend not to notice that the sun rises in a more northerly point in summer and a more southerly point in winter but this fact would have been well known to a people who lived much closer to Nature. The apparent 'wandering' path of the sun could have been a source of concern to a people who did not have our knowledge of the solar system. John Edwin Wood in his book *Sun Moon and Standing Stones*[1] shows how Stonehenge, with its extensive alignments with the summer and winter solstices and moonrises, could have been built as an astronomical observatory[1]. It could have been used by an elite group of astronomical scholars and astrologers to advise on the correct time to plant crops, the times of tides and auspicious times for events to take place.

Astronomical alignments were measured by taking sights using standing stones or even poles and lining them up with distant landmarks or other stones. This is illustrated in Figure 1 overleaf, and could all have been done in the open air without any need to go underground. Wood points out that while Newgrange can be used to tell the time of the winter solstice, this could be done more simply out of doors with a few standing stones. "There is no need to devise such a complicated arrangement", he says, "or to make it so large…perhaps the burial chamber received each year a sacrifice to ensure the lengthening of the days after the solstice".[2]

It seems that if the main purpose of Newgrange was not to measure the position of the sun; it might have been some ritual purpose related to it. While the course of the sun could have been tracked locally by the elders using the many standing stones dotted around the countryside, it is more likely that Newgrange was the centre for the thanksgiving ritual celebrated each winter and attended by large groups of people. But Newgrange must have been more than this. Why was it constructed with alternate layers of stone and turf? This feature is certainly not necessary for determining the time of the winter solstice nor does it seem to have had any ritual significance.

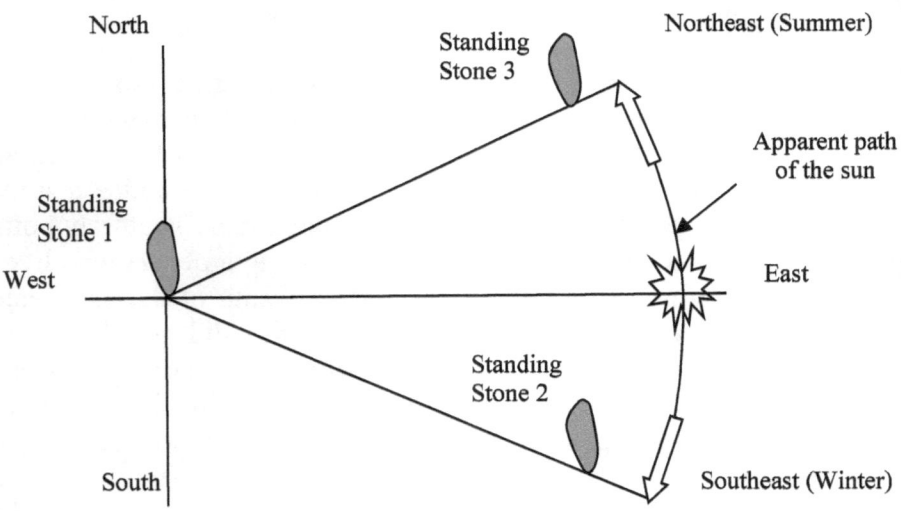

Figure 1. The apparent 'wandering' path of the sun can be checked above ground. Alignment of Stone 1 and Stone 3 with the horizon would mark the most northerly sunrise in mid-summer and alignment of Stone 1 and Stone 2 would mark the most southerly sunrise in mid-winter.

One kilometre to the west of Newgrange is the even larger mound of Knowth which turns out also to have alternate layers. These are described by Professor George Eogan in his book *Knowth and the Passage Tombs of Ireland.*[3] As mentioned in Chapter 3, Knowth has two passages facing east and west and aligned with the sunrise and sunset at the time of the equinoxes. Around Knowth are the remains of a further seventeen smaller satellite mounds with no discernable pattern of alignments.

The alternate layers of Knowth differ from those of Newgrange. Knowth's layers are made up of sods of turf, stones, boulder clay and shale. Boulder clay is a deposit of clay containing boulders which is formed by retreating glaciers. Shale is a type of rock formed by the compaction of clay and mud and can have significant amounts of carbon in it. The fact that the layers alternate following a fixed pattern of materials cannot be ignored. There must have been a good reason for it.

The archaeologist Laurence Flanagan in his book *Ancient Ireland: Life Before the Celts*, suggests that the sods of turf in Newgrange had been used to hold the stones in place.[4] It is possible that the layers in Knowth were also used for this purpose but this does not explain their complexity. In constructional terms there are a number of points to consider. In building a mound using locally available loose stones gathered from the surrounding area it is possible that many of these stones will be weathered and somewhat rounded as in the case of Newgrange and Knowth. These may have a tendency to roll and so it may have been necessary to bed them down in sods as suggested by Flanagan. This would not have been the case in Knocknarea in County Sligo where the stones were quarried and carried to the top of the hill to make the cairn. However since Knocknarea has not been excavated, it is not known if it contains any layers and I have not been able to find any references to layers except in relation to the Boyne Valley mounds. Another way of keeping the stones in place is by the use of kerbstones, and it is clear also that this use of kerbstones had been established long before layers were used since they can be seen at the earlier Mound 51 in Carrowmore.

However there is another approach which may not have been considered by the archaeologists since they tend to concentrate on the various objects used by ancient peoples rather than on matters of construction or energy. Flanagan quotes a "self-confident" archaeologist who defined archaeology as "the study of social and economic history through the actual commerciable products of society". This rather outdated definition is admitted by Flanagan to be too narrow but his book still devotes most of its pages to the "grave goods" and other finds in Neolithic sites.

Archaeologists have not devoted any of their time to considering the energies of ancient sites. We know that energy is constantly being received from the sun in the form of photons and other particles. This energy in the form of torsion fields is able to penetrate into the mounds. We know from the construction of orgone accumulators that they have alternate layers of inorganic and organic material. We also know that electromagnetic waves do not penetrate far underground. For example, a car radio does not work in a tunnel. Thus the suggestion that Newgrange and other mounds were types of orgone accumulator charged with electromagnetic and torsion energy by the sun appears to be a more reasonable explanation than that of an astronomical observatory. The earth and stones screened out the electromagnetic energy leaving the torsion energy to charge up the alternate layers of organic material

Orgone accumulators have been shown to have healing properties, therefore Newgrange and the other mounds could have been ancient healing centres. We do not understand how these energies worked but we know that a person resting or meditating or chanting inside one of these mounds would have experienced a profound energy effect possibly caused by resonance with a torsion energy field. We have been told by Nachalov that the structure of the torsion field of every object can be changed by the influence of an external torsion field. It is possible that many of the megalithic sites around Ireland were local healing centres and that Newgrange represented the perfecting of a long programme of trial and error in finding out which materials worked best and in which combinations. The seventeen satellite mounds at Knowth may have been prototypes or mini healing mounds built to test different materials. Eventually it was found that the alternate layers of organic and inorganic materials were best and a massive project was undertaken to build Newgrange to this design.

So I suggest that this development programme started in Carrowmore in 3550 B.C. and consisted of a large dolmen with a 30 tonne capstone covered in a large mound of loose stones. Perhaps it was considered that a larger mound would be more effective so Knocknarea was built with 70,000 tonnes of stones. From here the development may have moved to Carrowkeel with its mountain top site in visual contact with Knocknarea. Here we see a large number of smaller cairns and the first sign of solar alignment.

When the features of the various sites are laid out in tabular form below some idea of the stages of development can be gained. It is clear that as one moves from Carrowmore to Newgrange, features such as solar alignment, layers of different kinds of materials and spiral designs on stones appear at various stages on this south-easterly journey.

Site	Date B.C.	Alignment	Layers	Spirals	Standing Stones
Carrowmore	c. 3,550	No	No	No	None
Knocknarea	c. 3,200	No	Not known	No	None
Carrowkeel	c. 3,000	Yes	No	No	None
Loughcrew	c. 3,200	Yes	No	A few	Single stones
Knowth	c. 3,200?	Yes	Yes	Some	Single stones
Newgrange	c. 3,200	Yes	Yes	Many	35-38 stones in a large circle

Table 1. Comparison of features of the sites from Carrowmore to Newgrange

It can also be seen that the dates given do not flow in a smooth sequence from Carrowmore which is the earliest to Newgrange as the latest. All dates going back this far have a margin of error so the dates should be considered approximate. It seems that these sites could all have been in simultaneous use and could have had visual communication with each other. Perhaps they functioned as part of the same system. They may have been part of an energy grid of leylines, a point which I will come to later.

Knowth is suggested by some people to be of about the same age as Newgrange while others consider it to be older. Eogan reports radiocarbon dates for both Newgrange and Knowth of which one from the base layer of Knowth is given as 4885 $\pm$ 110 b c (uncalibrated). This is much earlier than any of the dates given for Newgrange and I believe that Knowth is older than Newgrange and that it was its forerunner. My belief is that Knowth was the first site to experiment with the effects of alternate layers. The seventeen satellite mounds were possibly part of an experiment to test different arrangements of materials. Unfortunately many of these mounds were found to be in such a poor state that the existence of layers could not be confirmed. However, of the seventeen, Eogan describes five as having layers and two as possibles.[5]

The principal technique used to date these sites is radiocarbon dating. All living organisms absorb the radioactive isotope Carbon 14 in small quantities. When the organism dies, it ceases to ingest Carbon 14 and that which is in it begins to decay. A measurement technique is then used to see how much of the Carbon 14 is still present. Since the rate of decay of this isotope is known, a calculation of the amount of Carbon 14 will indicate the age of the material. Newer material will have lost less Carbon 14 than older organic matter. But dates can be inaccurate for a number of reasons. Apart from the accuracy of the estimates of exposure to Carbon 14 over the ages, there is also the problem of whether the sample is contemporaneous with the site. In many cases older material has been used for construction or more recent settlement has introduced newer material. There are also a number of other dating techniques and they do not all give the same results so we have to be careful in comparing dates of different sites.

From the table it can be seen that solar alignments begin to appear at Carrowkeel and can be found in all sites as we progress from there through Loughcrew and Knowth to Newgrange. In considering the developments from the energetic aspect, it is only in Knowth and then in Newgrange that alternate layers begin to appear. The earliest sites seem to have relied on large quantities of stones without any intermediate layers of material. Perhaps the earlier sites were developed as far as possible using larger masses of stones and it was then discovered by psychic means or by experiment that the energy of the site could be enhanced by alternating the stones with these other materials. If we accept for a moment that energies could have been visualised by these people we begin to have some insight into the energetic aspects of these sites.

These ancient sites are considered by many people to be connected to each other by a massive subtle energy grid. The lines that connect them, the leylines as they are called, are based on the discoveries of a psychic named Andrew Watkins. He claimed that in 1921 he had a vision of the leylines connecting well known sites in England. They went from hilltop to hilltop and took in ancient sites, standing stones, holy wells, churches and crossroads. He and many others after him have sought out these alignments and their work is described, as are the many alignments, in *Earth Magic* by Francis Hitching.[6]

Hitching, though not a scientist, writes with a degree of scientific rigour which belies the subject of his book as suggested by its title. He describes the measurement of magnetic fields around standing stones which are many times greater than can be explained by any natural variations. These stones always seem to be located above underground springs. These are what water diviners call "blind springs" where water rises vertically from far below but is unable to break the surface and radiates out horizontally in one or more underground streams.

The work and theories of Bill Lewis, a talented diviner and electrical engineer, are quoted at some length by Hitching.

> Lewis, through his work as an electrical engineer, says that there is experimental evidence which shows that the movement of water through a tunnel of earth, particularly in clay, creates a small static electric field. The crossing of streams, albeit in different strata, makes the field stronger. He believes that the stone, placed immediately above this, acts in some way as an amplifier, although there is no known theory of physics to explain how.[7]

Lewis is also quoted as stating that "the power when it emerged from the ground and up the stone, came in the form of a spiral". This force varied and even changed polarity reflecting astronomical changes such as the phases of the Moon.

I was puzzled by the fact that the leylines run through places where churches had been built until I read Hitching's explanation that in the early years of Christianity, churches were built on top of old pagan sites of worship. Whether this was recognition of the energy field of a sacred place, or whether it was an attempt to stamp out the ancient pagan practices is not clear. However, it seems that the early church leaders decided that it was easier to absorb the old religion than to defeat it.

The purpose of leylines is a subject of debate even among those who recognise them. Some people say that the system of leylines is an energetic grid through which subtle energy flows to regenerate parts of the planet. Others think that leylines may have been used for navigation purposes or that they are energy fields laid down by the passage of animals or humans over many years. Those who follow the energy grid theory believe that subtle energies enter the earth through major energy centres such as Newgrange, Stonehenge and Avebury and are distributed through the leylines to the rest of the countryside.

Many people refer to alignments between major Neolithic sites but these do not seem to be confirmed when drawn using the correct methods. There seems to be some alignment between Carrowmore, Newgrange and Avebury as seen in Figure 3. but it is not a straight line as some people claim and it does not pass through Stonehenge. In drawing such diagrams it is important to use a map which views the world as a globe as has been done in Figure 3. Ordinary maps from atlases will introduce errors over large distances because they are normally based on a cylindrical projection which flattens out the globe and produces distortion at higher latitudes.

Another insight into the origin of leylines may be obtained by considering the Aboriginal concept of songlines in Australia. This very ancient people regards the songlines as hidden roads that enable them to find their way across the continent without maps. No matter where the individual is, the starting point has a song that leads to the next stop. In going from one place to another the aborigines are said to 'sing the world into existence'. They create the world through their consciousness as they move along.[8]

We can also see from the table that standing stones appear for the first time at Loughcrew and then at Knowth although these are rather small.

While stone circles have been seen in the earlier constructions such as Carrowmore, these could not be considered to be standing stones.

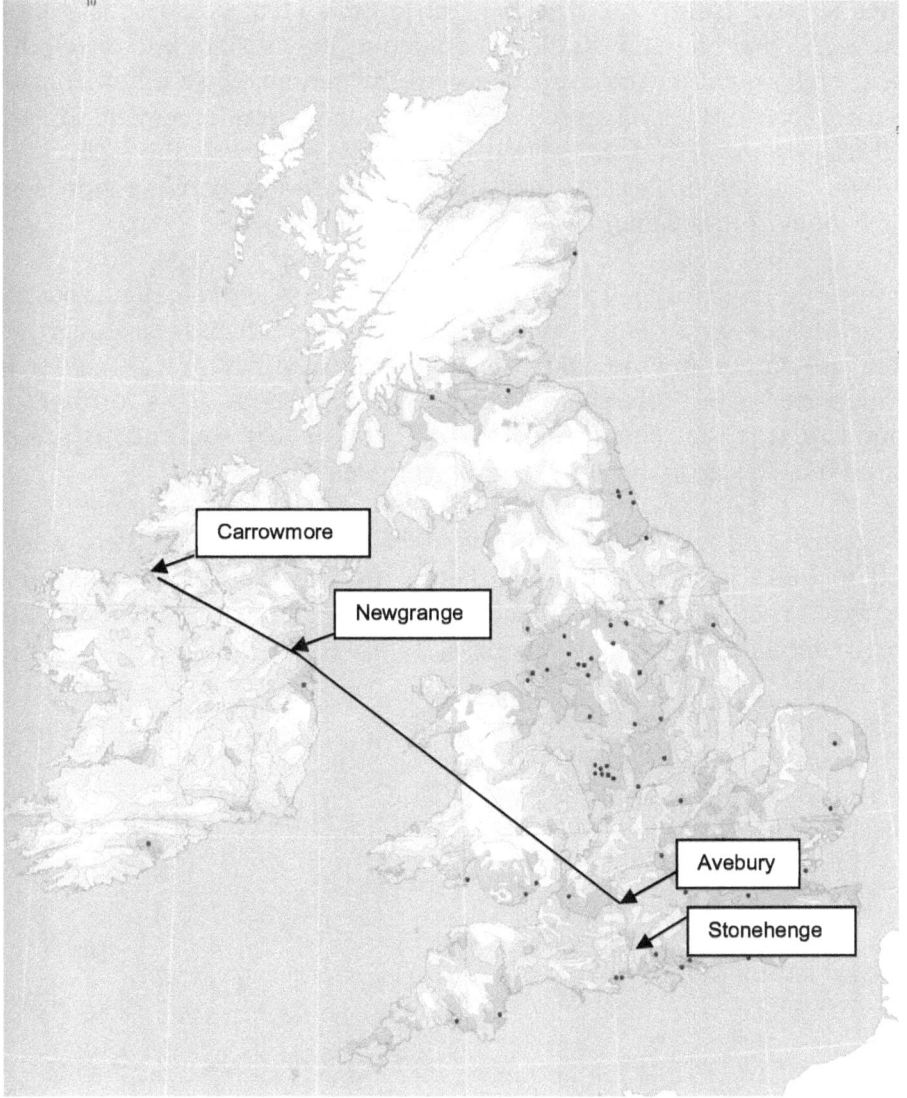

Figure 3. Possible energy grid leyline between Carrowmore, Newgrange and Avebury. This does not pass through Stonehenge. Standing stones become more common and more impressive as we move southeast.

One of the notable characteristics of standing stones is that they always stand alone and are never placed adjacent to other standing stones. There is evidence of a magnificent standing stone circle at Newgrange. Twelve stones survive from what was thought to be a great circle consisting of between 35 and 38 in all. Radiocarbon dating shows that these were put in place some time after the construction of the mound so we might consider them to be a further upgrade, a possible energy reinforcement of some kind. As we move further beyond Newgrange and cross the Irish Sea to Avebury we find one of the most spectacular sites of standing stones and to the south of it, the later construction at Stonehenge.

While some have argued that the construction of the so-called "passage graves" began in the Boyne Valley and spread westwards to Loughcrew and Carrowmore, my view that these structures evolved from west to east is supported by Professor O'Kelly who says that it is difficult to accurately date the sequence and "if an evolutionary sequence is present at all it must be from the simple to the complex".[9]

Newgrange was more than an astronomical observatory and there was no need to build such a complex structure to find the date of the winter solstice. It is clear that something else was going on which involved some form of subtle energy which was possibly used for healing.

Chapter Nine

Newgrange is Damaged

It is believed that Newgrange was used for ritual purposes but I have provided some evidence that it also performed the function of an energy concentrator. This may have benefited the surrounding area, and may also have been used for the purpose of healing or for divination. It appears that Newgrange with its alternating layers of organic and inorganic material was able to function as an orgone accumulator or a concentrator of torsion fields. The question is, can it still do so?

I would like to see a research project carried out which would measure the effects of spending time in the chamber at Newgrange. It would be interesting to see if any healing or energy effects were experienced. Regrettably the results may be disappointing, as a number of attempts have been made at restoration over the years. Professor O'Kelly reports in his book on Newgrange that modern materials such as concrete and plastic have been used. Concrete pillars were installed and a concrete cover built over the passage and chamber. A concrete wall was added behind the kerbstones around the outer edge to stabilise the structure. Figure 1 overleaf, is a photograph taken looking east towards the entrance and from above the passage, showing the concrete cover over the passage and the new retaining wall that was built at the entrance.[1]

This "restoration" of Newgrange was carried out with the best of intentions but using the knowledge available at the time. Excellent engineering and architectural work was undertaken and many technical problems were successfully addressed. However, the possibility that Newgrange was an energy device was not taken into account and the main considerations were for the stabilisation of the structure in order to make it safe for tourists to visit.

Perhaps Newgrange can no longer function as originally intended.

When Professor O'Kelly started his excavations in 1962 the mound was in a sorry state. Years of neglect, the carting away of stones for building materials, the planting of trees and some reconstruction in the 19th and early 20th centuries had all taken their toll. In addition while the construction methods of the Neolithic builders were excellent for their time, the building could hardly have been expected to last 5,000 years. Much of the cairn material had slipped from its original positions and many of the kerbstones were tilted forward or pushed over. One of the roof slabs in the passageway was broken and some of the orthostats (vertical stones) holding up the roof slabs were out of position. Some of the stones in the corbelled roof of the chamber had cracked and water was seeping in. Concern had been expressed for the stability of the structure and some ill-advised restoration work had been done in the chamber by erecting concrete pillars some of which had to be removed.

Figure 1. The "restoration" of Newgrange using concrete and other modern materials. *(Reproduced with permission from the estate of Michael J O'Kelly)*

Figure 2 opposite illustrates the restoration work which was carried out after Professor O'Kelly's excavations finished in 1975. This shows the extensive reinforced concrete construction undertaken. The upper drawing shows a cross-section of the mound with a cover constructed over the passageway. This cover has a manhole so that archaeologists and maintenance staff can gain access to the top of the passageway. Also illustrated is the "cowl" or internal roof over the chamber. This can be seen in end view at the lower right hand side. The middle section of the drawing shows a plan view and the modifications to the entrance as seen in Figure 1. Dotted lines show the "probable line of original quartz/granite" which can be compared with the revised entrance, widened to allow access for visitors.

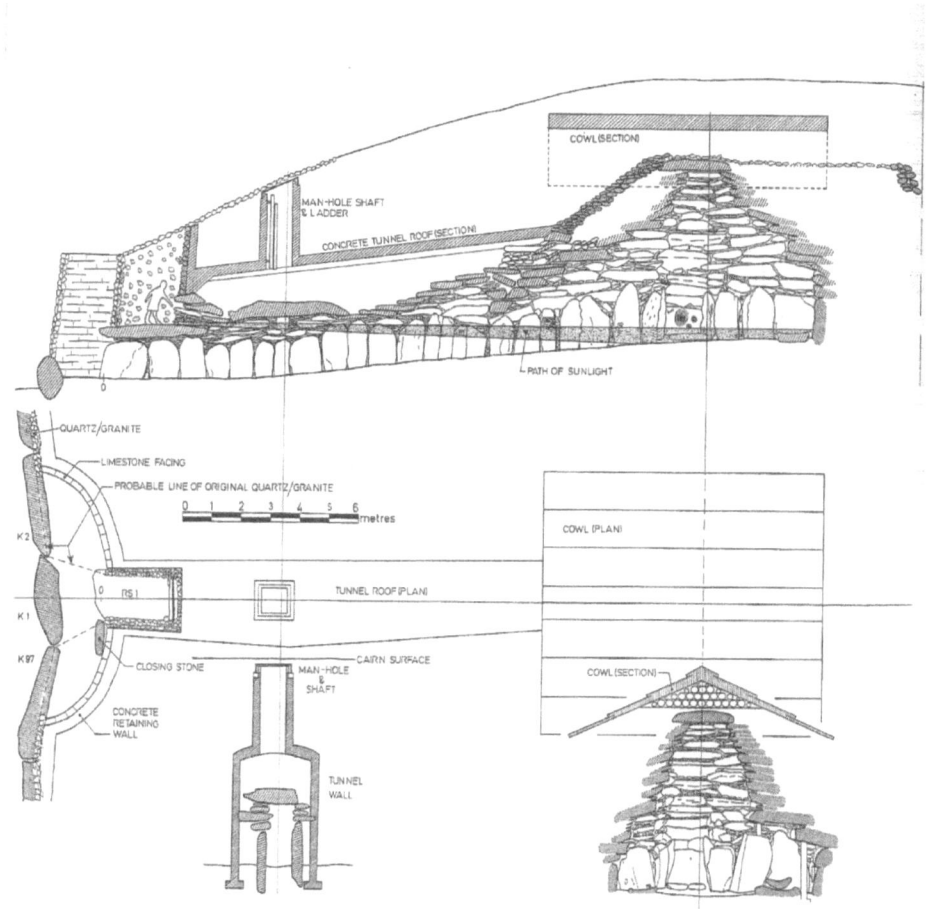

Figure 2. Plan view and sections through the mound showing the concrete tunnel over the passage, the path of the sunlight, the manhole access shaft and the cowl over the chamber. *(Reproduced with permission from the estate of Michael J O'Kelly)*

79

The widened entrance has been the subject of controversy as has the vertical retaining wall with its facing of white quartz. It is clear from the excavations that the quartz was an external feature but it is by no means clear that it formed a vertical wall as can be seen in Figure 5 of Chapter 1. This is referred to by the archaeologist Geraldine Stout in her excellent work on the history, geology and archaeology of the Newgrange area, *Newgrange and the Bend of the Boyne*. "The white quartz wall has a glaringly modern appearance", she says, "….it is difficult to imagine that the monument could ever have appeared like this in prehistoric times. Its construction has to be understood in the context of the late 1960s fashions in restoration. Many have argued for its removal".[2] In a more recent book Stout describes this as a "monumental mistake" and states that "it has been included in an international list of the world's worst archaeological reconstructions".[3]

To counter the ingress of water a drainage system was installed and a concrete wall was built behind the kerbstones to protect them and save them from carrying the weight of the loose cairn material. Plastic sheeting was used to make sure the concrete did not adhere to the kerbstones. When the cairn material was replaced, heavy plastic sheeting was placed over the layer of stones to prevent soil washing down into the drainage system.

Today's visitor to Brú Na Bóinne, the area surrounding Newgrange, is directed first to an impressive interpretive centre on the southern bank of the Boyne. Guided tours are laid on at fixed times to both Newgrange and Knowth, the whole visit taking about three hours. Visitors cross the Boyne by a pedestrian bridge to a collection point from which they are brought in groups by minibuses to the adjacent sites. Visitors to Newgrange are brought inside the mound and I described in Chapter 3 how the winter solstice sunlight is simulated by shining a light through the entrance.

The designation of Brú Na Bóinne, as a World Heritage Site by the United Nations Educational Scientific and Cultural Organisation UNESCO in 1993 has brought with it significant obligations to preserve the monuments of Newgrange, Knowth and the lesser known mound of Dowth, and to make them accessible for visitors. However, it is clear that significant pressure existed even before this which resulted in the modification of the entrance to Newgrange to allow the smooth flow of visitors to the chamber.

While the restoration of Newgrange has been well documented in Professor O'Kelly's book, the more recent restoration of Knowth has not been so well described. I visited Brú Na Bóinne in April 2007 with a view to seeing the present state of the monument for myself. The restoration of Knowth has been the subject of some discussion as it has been reported that the eastern passage had been blocked by the construction of a concrete wall thereby preventing the entrance of the sunlight at sunrise on the equinox.[4]

Figure 3. Entrance to the concrete room showing the concrete wall (on right), blocking the eastern passage at Knowth. *(Photo K. Comerford)*

My visit to Knowth took an hour and a half in all and the guide was very professional and well informed. Now that the restoration of Knowth is complete, the site is fully accessible to visitors and I was particularly interested to see what had been done to the entrance to the east passage. I was surprised to find that not only had a concrete wall been erected but a concrete room had been built inside the mound. This is presumably to try to give visitors some experience of entering the mound as otherwise they might be disappointed especially if they had visited Newgrange first. The two passages at Knowth have very low ceilings and it would never have been possible to allow people to enter them.

The concrete room has been hollowed out of the layered structure of the mound. It measures about three metres by ten and the outermost end looks down into a pit allowing visitors to see daylight through a grill located behind two of the kerbstones. Inside the room are illuminated wall panels which show the constructional details of the mound and the interesting artefacts found inside it. The most famous of these is the so-called "mace head", one of a number of carved items with holes in them found in the mounds and which have been described by diviners as pendulums for dowsing although archaeologists prefer to refer to the smaller ones as pendants.

The concrete wall and narrow door at the eastern side of Knowth can be seen in Figure 3 above. Access is gained by walking over a small footbridge and down the steps shown in the photograph. On entering the mound the first thing that one sees is a souterrain or underground storage tunnel dating from the time of early Christian occupation of the site. Further on one can see the main passageway which is 35 metres long and ends in a cruciform chamber. This is closed off by an iron gate and it is clear that you could only enter it on your hands and knees, not an ideal situation for tourists! To the left one enters the concrete room which has no real connection with context of the mound and serves no apparent purpose other than to give a feeling of having entered the mound and a place for the guide to explain its features.

On continuing the tour outside, one sees the many decorated kerbstones for which Knowth is justly famous. These are now protected by an overhanging concrete ledge. Passing the southern side of the monument I was most surprised to see that a ramp had been constructed to facilitate tourists who wished to walk to the top of the mound to enjoy the view and take photographs. This "tourist trail" which slices through the layers of mound material can be seen in Figure 4 opposite.

Professor O'Kelly reports that 70,000 people visited Newgrange in 1978. By 2006 this number had reached 233,509. It is clear that tourism interests have overridden the wishes of those who want to see the sites properly preserved and with the least modification. A petition was organised by a group called Global Vision who wanted the Irish government to remove the concrete wall and room from Knowth, to suspend any further restoration work and only carry out preservation using original materials and methods. They also expressed concern for the damage being done to Newgrange by the large number of visitors entering the chamber.

Large amounts of the layered materials of Newgrange and Knowth have been removed as part of their restoration. In their place we find modern materials such as concrete and plastic. If these sites were energy devices depending on their layered construction to function, it is clear that they can no longer fulfil this purpose. However, archaeologists are meticulous in documenting everything they find or have to disturb. It is hoped that sufficient records remain to reconstruct these monuments to their original state if, in the fullness of time, the lessons which Newgrange can teach us are fully understood.

Figure 4. The "tourist trail" leading to the top of Knowth. The concrete protective ledge over the kerbstones can also be clearly seen.
(Photo K. Comerford)

Chapter Ten

The Lessons of Newgrange

There is much that can be learned from Newgrange if one approaches it with an open mind. It is not that the traditional views are wrong, just that they are narrow in their focus and approach. Most of what we know has been told to us by archaeologists who have made their discoveries and come to their conclusions using their traditional methods and practices. These are to excavate a site and to find and identify objects which reveal something of the people who lived there. Archaeology is a branch of anthropology and is therefore concerned mainly with people. Newgrange seems to have been about much more than people and perhaps we have to look outside traditional archaeological practices and sources to fully understand what was going on there. Let us therefore review what we have learned so far and see if the ideas introduced to date can lead us to a broader understanding which might have applications in the 21st century.

It is a somewhat narrow view to describe Newgrange as a tomb and many modern archaeologists are moving towards the broader description of these monuments as mounds or cairns. Cremated bones were found in a number of these structures and if we look at the practices of early peoples we find that cremation took place outside the structure and it appears that bones were brought inside for some ritual purpose. Thus the cremated bones could be remains of people who had died and were being venerated or perhaps ceremonies were being carried out to help their spirits move on to the next level in the spirit world. To understand these activities we must look beyond the traditional preoccupation with burial and grave goods and try to understand something about the beliefs and spiritual practices of these people.

A number of researchers in different disciplines have produced convincing arguments that a peaceful society existed prior to 2,000 B.C. Riane Eisler based her work on the recent discovery of the Minoan civilisation. James DeMeo's work is based on the detailed study of climate change in the region he calls "Saharasia" which led to competition for scarce resources and the beginnings of the use of physical force. Nick Brooks of the University of East Anglia has argued that civilisation as we know it today resulted from the need to reorganise in order to meet deteriorating climatic conditions. There is evidence that the Sphinx was once in a temperate region and became weathered by rainfall. Jonathan Haas of the Chicago Field Museum has not been able to find any evidence that his warfare hypothesis could explain the beginning of organised societies. Finally, there is no sign of any kind of military or defensive construction in Ireland prior to 2,000 B.C.

There is a folk memory in many cultures which speaks of a utopian society which existed long ago. Christian, Jewish and Islamic traditions refer to the "Garden of Eden". Buddhists refer to "Nirvana" and "Shambala", while in India the Vedic civilisation is claimed to have experienced "Heaven on Earth". If we accept that a peace-loving people existed in Ireland and elsewhere prior to 2,000 B.C. and wish to investigate this subject with an open mind, we must step back from our current set of values based as they are on private property, competition in business and personal security. Only then can we have any possibility of understanding the way of life of these prehistoric societies. Like many indigenous peoples they probably had no concept of land ownership or private property and had a strong sense of community. It is likely that they suffered much less from stress than we do and thus their nervous systems may have been much calmer and more settled than ours. This could have increased their sensitivity to subtle influences and energies and given them a highly developed intuitive sense.

The spiral shows us an aspect of Newgrange which has not been explained by archaeologists or anthropologists. There are a number of possible explanations of its significance. From a spiritual point of view it is seen as representing the inner journey from the manifest world to the unmanifest. In Newgrange it could represent the vision quest of the shaman or most psychic member of the community. This could be assisted by drumming, chanting, by psycho-active plants or just by sensory deprivation. There are also a number of physical explanations. The spiral is commonly found in Nature, for example in the growth of plants, in hurricanes and tornadoes and in galaxies. It is also called a vortex.

Vortices can contain a lot of energy and because angular momentum is conserved in a spinning substance, this energy decays very slowly. Thus, spinning substances, such as Lord Kelvin's smoke rings, represent a stable form of energy. The rotational speed of a spinning substance can be described in terms of frequency and substances vibrating with similar frequencies can resonate with each other and transfer energy.

Dr. Valerie Hunt, Founder of the Bioenergy Fields Foundation has been researching bioenergy, the energy of living organisms, for over fifty years. She started her research on human energy fields in the 1960s but found that the instruments available at that time could not detect frequencies above 150 Hz. However she found that NASA had used higher frequency instruments to record signals from the brains, hearts and muscles of astronauts. She approached the manufacturers of these instruments and with their help she has been able to measure bodily frequencies up to 750,000 Hz.[1]

By testing different individuals she has found that the frequency of the signals from the average person peaks at about 250 Hz which is the same as the frequency of signals from heart and muscles. However, for people with healing and psychic ability, a higher frequency signal was detected which peaked at 800 Hz. Hunt then found that people who were evolved spiritually, such as mystics, had frequencies up to 200,000 Hz. She says that we need a new model of understanding of the human energy field based on spirituality or consciousness. From this we might conclude that the spiral, a representation of energy, possibly at a higher frequency, could have been perceived by the ordinary people of Newgrange in a manner in which only psychics can see today.

We are learning that, when we look at spiral energy, there is a physical dimension, but there is also a metaphysical dimension, i.e., one that is beyond physics. That is because some aspects of this energy have been shown to exist which are not detectable by conventional instruments. One example of this is the torsion fields described in Chapter 4 and which are still the subject of debate among scientists. These spiral energy fields have unusual properties. They are related to electromagnetic fields, and just as a substance can be magnetised by an electromagnetic field and its structure permanently changed, torsion fields can also "magnetise" a substance and change its structure and they could be another name for the orgone energy discovered by Wilhelm Reich. This could be the basis of healing using subtle energies.

Russian scientists investigating torsion fields also found a number of anomalies in the way spinning objects appear to defy the law of gravity. The famous British engineer Professor Eric Laithwaite demonstrated that a rotating wheel becomes lighter. Professor Myron Evans has produced a theory of torsion fields suggesting that gravity may be overcome and which is attracting serious attention. Further research on these subjects may lead to an explanation of how the heavy stones of Newgrange and Stonehenge and the *Grand Menhir Brisé* ("Great Broken Menhir") at Locmariaquer, Brittany were lifted into position. Even with the most modern construction technology we cannot even begin to understand how the Great Pyramid of Giza was constructed. Christopher Dunn tells us that it would take 1,800 men to haul each one of the 2.3 million blocks of stone into position.

Unfortunately, scientists and engineers who try to come up with possible explanations for this form of energy face stiff resistance from traditionalists who are reluctant to accept any explanation until it has been fully proven and is seen to accord with existing theories. These ideas are labelled as "pseudo-science" but so were many new theories in the past until a full explanation of them was found. How is it that many formerly respected people such as Professors Laithwaite and Evans are suddenly shunned as soon as they propound a theory which appears to contradict the establishment? In my view this has more to do with attitudes than with science. What is clear is that there is much that we do not know. There appears to be a form of energy as yet not fully understood but which could have enormous benefits for humankind. We may need to suspend our scepticism if we are to give this energy a chance to emerge. Richard Feynman the Nobel Prize-winning physicist stated that we must frankly admit what we do not know. He said that "doubt and discussion are essential for progress into the unknown" and went on to say that it is the responsibility of scientists "to teach how doubt is not to be feared but welcomed and discussed; and to demand this freedom as our duty to all coming generations".[2]

The theory of earth crust displacement is not conclusively proven but there are many indicators that it may have happened. One of these is the definite evidence that the climates of some regions were different in the past. This may have led to the migration of peoples and to competition for food and water, which may have triggered the end of the partnership society. The folk memory of the sun changing its position in the sky may explain the preoccupation of Stone Age peoples with solar alignments. It may also explain the stone rows at Carnac in Brittany and the alignment of Newgrange with the winter solstice. These may have been used to reassure the people that the sun was halting its southerly drift and moving back to its previous position.

Subtle energy in the form of torsion fields may be attracted to organic material and alternate layers of organic and inorganic material may be able to amplify and store this energy. Wilhelm Reich claimed that orgone energy had healing properties and that it was negatively entropic. This means that it had the effect of reversing the natural tendency of objects to decay and that it had organising power. The discovery by Professor O'Kelly that Newgrange has alternate layers of organic and inorganic material leads one to suspect that Newgrange operated as an orgone accumulator and possibly had healing properties.

We might be able to have more confidence that Newgrange was a healing device if more rigorous research had been done on orgone and had been published in peer-reviewed journals. Regrettably because of the harsh judgement meted out to Reich by the US courts and the burning of his books by the Food and Drug Administration, including the banning of work on orgone, it has not been a respectable research subject. In spite of this James DeMeo and others have continued to work in this area and he lists seventy research papers in his *Orgone Accumulator Handbook*.[3] In this he gives detailed instructions for constructing a range of orgone devices and relates how they have been used to improve weather conditions and to cure burns, cuts, cancer, herpes and arthritis.

I believe that the principal lesson we can learn about the purpose of Newgrange is that it was a device used for activities which are not accepted by science today. Consequently these activities are not considered or understood by today's researchers. I believe that researchers must look beyond Newgrange, beyond the more obvious physical aspects of it to see where such enquiries may lead us. The may lead us to a "New Science" one which has room to accept new ideas and new methods. I have shown that Newgrange represented the culmination of a development project where stone structures were gradually refined in their construction, leading to the possibility that they became highly effective places for healing and divination. If this is true we must ask what we can learn and how we can benefit from this knowledge.

The existence of Newgrange demands that we ask what form of energy was associated with these activities and if we do not understand it, we must, as Richard Feynman advised, admit what we do not know and not be afraid to face our doubts. The principal lesson we can learn about the energy of Newgrange is that it has qualities which go beyond the known physical realm, beyond the five senses. The partnership society, with its calmer way of life, may have enabled the people of Newgrange to function in a much more intuitive way than we do today. To pursue this investigation further we need to understand the way of life and outlook of the Newgrange inhabitants and even the states of consciousness that they experienced. This is not easy to do from our 21st century viewpoint but one approach is to study the nature of shamanic practice which was used in Newgrange and which has recently been revived and which must cause us to change our ideas about reality.

Chapter Eleven

The Revival of Shamanism

Shamanism is an ancient spiritual system whose roots go back before records began. It is claimed to be the forerunner of all spiritual systems that we know today. Ancient tribes and even some surviving to this day, usually have a shaman or witch doctor who is considered to have special powers of healing and divination. This individual attempts to enter an altered state of consciousness in order to obtain knowledge and power, to contact spirit beings or to perform healing. This is done by the use of psycho-active plants, by drumming, dancing or through the dream state. A shaman always works in close contact with nature and is considered to be able to recognise and work with the elemental forces. From a scientific perspective we could say that the shaman has developed an ability to access altered states of consciousness or to work with subtle energies.

In recent times there has been a revival of interest in shamanism, no doubt associated with the increased interest in older spiritual knowledge and systems. No longer is shamanism associated with the witch doctor image. The shaman is one who works for the benefit his or her community and not to the detriment of any other person. It is now possible to explore many of the aspects of shamanism as a modern spiritual practice. The Irish Centre for Shamanic Studies has been established close to the River Boyne only a few miles upstream from Newgrange and provides a range of courses and workshops.[1]

In a shamanic journey or vision quest, an ordinary person can be guided by a shaman to seek answers to questions that arise concerning the conduct of one's life, blockages relating to the past or health issues. All of the shamans that I have met have been trained by Native Americans. They use natural techniques such as dance, drumming and music and often create atmosphere by burning dried sage and wafting the smoke around people using a large feather or bird's wing. People can have 'out of body' experiences which can be very significant in meaning. Usually one is directed to visit the world below to contact an animal spirit or the world above to contact a human spirit.

In one journey I undertook, I was put in a reclining position in a chair and listened to a tape of monotonous and insistent drumming. I was directed to look for my power animal which can be found in the world below. After some time I became aware of being on the side of a steep hill. It was a grassy slope leading down to a rushing mountain stream. All around were tall pine trees and beside me was a vixen and her cubs. She was my power animal and I christened her "Vicky". We went for a run down to the river but instead of taking some time to get there, we were there instantly splashing around in the water. I believe that this experience was telling me to let go, to stop taking myself so seriously and to focus on the present moment.

After some time Vicky brought me to the world above where a wise being communicated with me telepathically. I could not see this person but I received a number of insights concerning events in my life and a message which I believe was from my father who had died when I was a child.

The shaman was the most psychically gifted person in the tribe and carried a huge responsibility. If danger or illness threatened, the shaman was expected to be able to find the answers. If he did not find good enough answers he would lose respect. His power came not from the use of physical strength but from subtle energy. If he could not command this energy he had to rely on fear. It is clear that some lesser talented shamans used fear and superstition to survive but this is seen in the more primitive societies. I use the word primitive here in one of its dictionary meanings which is "crude or simple". The other dictionary meaning is "of the earliest times" and does not include any suggestion of lack of intelligence or culture. We must be careful how we judge people from societies which are different to ours.

While much is talked about native tribesmen engaging in practices that seem ridiculous to us, it is wise to suspend judgement until we can experience the world of the shaman and understand the forces and energies being brought into play. While we may not believe that a certain shaman is genuine, we must remember that he is creating a reality which is true for his tribe. This limited reality, valid for them, contains the solutions to their problems. Without understanding the shaman's reality we should not form any judgements about it.

In understanding shamanism we are asked to consider a situation in which non-ordinary states of consciousness form part of everyday reality. Stanislav Grof has made an extensive study of these states, initially using LSD and later a form of hyperventilation or deep breathing which he calls Holotropic Breathwork.

Grof describes how as a young psychoanalyst in Czechoslovakia he became disillusioned with Freudian techniques which, while giving convincing explanations of mental states, did not seem to provide clinical solutions. When he received some samples of LSD he decided to test it and carefully record his experiences under laboratory conditions. He states:

> My first LSD session radically changed both my personal and professional life. I experienced an extraordinary encounter with my unconscious, and this experience instantly overshadowed all my previous interest in Freudian psychoanalysis. I was treated to a fantastic display of colourful visions, some abstract and some geometrical, others filled with symbolic meaning. I felt an array of emotions of an intensity I had never dreamed possible.[2]

From studying the reports of thousands of people who have experienced altered states of consciousness through Holotropic Breathwork, Grof states that "we are not just highly evolved animals with biological computers embedded in our skulls: instead we are also fields of consciousness without limits, transcending time, space, matter and linear causality".[3] He describes different experiences which are common to large numbers of people. These include, going back in time to infancy or birth or to previous lives or experiences of in-between states similar to the 'bardo' states described in the *Tibetan Book of the Dead*. These are out of body states where people report visiting places in this and other worlds.

He goes on to describe what he calls the "transpersonal nature of consciousness". Here he explains that we have to move beyond the belief that our consciousness is limited to our brains and limited in space and time. It means accepting that our lives are not only influenced by our immediate environment but also, and equally importantly, by the ancestral, cultural, spiritual and cosmic influences beyond our normal sensory range.[4]

In carrying out his researches he has acknowledged the shamanic traditions and has worked with a number of modern shamans. From his studies of over 20,000 breathwork sessions with people from all over the world he reports that many ordinary people have shamanic experiences. They meet power animals, spirit guides and higher beings. At times people have reported visits to other realities that exist parallel to our own and experiences of a mythological nature. These relate to the archetypes described by the famous Swiss psychiatrist Carl Jung as manifestations of fundamental organising principles in the cosmos. People are experiencing the field "in which the devas responsible for the cosmos reside" as stated in the Rig Veda.[5] These experiences are universal and cross historical, geographical and cultural boundaries.

Michael Harner, author of *The Way of the Shaman*, an anthropologist who underwent a shamanic initiation in South America, explains that Western scientists can never understand the shamanic experience since their view is enthnocentric, based on their own culture and regarding other traditions as naïve, inferior and primitive. He also describes their approach as cognocentric, an academic approach that only recognises information received by the five senses in the ordinary state of consciousness.[6]

To the shaman, matter and energy are one. The vibrations of all things become known; the healing energies of plants, the energy of the sun or moon or of a particular time of day or night. Ross Heaven in his book *The Journey to You: A Shaman's Path to Empowerment* says that one definition of a shaman is a transformer of energy.

> Shamanic techniques provide for change by transforming our own energy and the energy of the universe we are part of, allowing us to blend, channel and direct it into the areas of our lives where it will have most beneficial effect for us and, through our connection to all living things, to the potential unfolding within the universe right now.[7]

Energy is understood by the shaman as vibration. By resonating with the different frequencies of the various plants, animals and even rocks, the shaman transcends the physical world as we know it. He or she believes that all levels of existence, including material and non-material levels, such as thoughts or feelings, have a certain vibration. If you can reproduce the vibration of the element you are working with, you can enter into it and transform it. This can include healing and it explains how some of the native healings may be carried out. By working at a non-physical level, subtle energy is used to transform the etheric or subtle body and this transformation manifests as healing in the physical body.

Edith Turner is a field associate of the Foundation for Shamanic Studies founded by Michael Harner. She is a distinguished anthropologist who teaches at the University of Virginia. She says of a native healing she witnessed:

> In a book entitled *Experiencing Ritual*,[8] I describe exactly how this curative ritual reached its climax, including how I myself was involved in it; how the traditional doctor bent down amid the singing and drumming to extract the harmful spirit; and how I saw with my own eyes a large, gray blob of something like plasma emerge from the sick woman's back.[9]

These are the words of a respected academic. Whether you regard the energy as spirit or subtle energy is really a question of the framework from which you are operating. From the framework of transpersonal psychology it is a harmful spirit while from a more technical perspective it may be called a negative energy.

The shamanic tradition is universal and very ancient. Many cave drawings have been found which suggest visions seen in trance states.[10] The stone decorations at Newgrange could easily be such. While a lot of effort has been devoted to interpreting the symbols at Newgrange and other mounds in physical terms, it is also possible to see them as representing non-physical realms or energies.[11] Perhaps they do not represent stars or calendars but spirit beings, archetypes or subtle energy flows. If we accept that certain people at Newgrange in Neolithic times experienced non-ordinary states of consciousness it is easier to understand how healing of the body took place there.

The shamanic tradition also works with Nature. The laws of Nature are considered to be overseen by the devas or elementals. From the five-sensory perspective these do not exist but to the multi-sensory person they are real, and the subjective experiences of large numbers of people cannot be dismissed except from a five-sensory or cognocentric viewpoint. The universal lore on subtle energy and elementals, found in all traditions but perhaps using different terminology, describes Nature as having four elements, earth, air, fire and water. There are elementals charged with overseeing each of the four elements.

The earth elementals are known as gnomes. They are concerned with movements of the earth and with mining and metals. The sylphs are the air elementals. They look after the winds and the weather. The fire elementals are called salamanders and they control the energy of combustion. The water elementals are called undines. They are in charge of the seas, rivers and lakes. Included in this tradition are the beliefs that these elementals should be respected and that serious consequences can result from undertaking insensitive actions which may affect them. This view implies that humankind is behaving with scant regard for the planet and that angry elementals can punish us by causing earthquakes, bad weather and other disasters. It is interesting to note that all of these have increased in frequency in recent years.

Other elementals are said to be in charge of the growth of plants, animals and humans. These are nature spirits which are concerned with plants and body elementals who look after creatures with bodies. Each human is considered to have a body elemental. One method of healing is to establish communication with this spirit and attune to it. In this way a person will intuitively sense when they are engaging in any action which might benefit or harm the body. Other methods consist of tuning-in to plants to learn how to use them for healing.

Now that we have some idea of the multi-sensory world of the Newgrange inhabitants we can consider how their knowledge of subtle energies was applied to other more physical activities. One aspect which has received little consideration from archaeologists is the importance of water. Many stone circles have avenues which lead to a nearby river. Examples of this are Stonehenge, Avebury and Stanton Drew in Avon. As well as rituals worshipping the sun and heavenly bodies there is evidence of water worship going back to 6,000 B.C.[12] It is likely that an avenue ran from Newgrange down to the Boyne although it has probably been ploughed over. This would have formed part of the type of rituals such as the Son et Lumière proposed by Hugh Kearns which I described in Chapter 3. He claims that the sunlight was reflected from Newgrange onto the surface of the Boyne and viewed by throngs of people standing on the natural amphitheatre provided by the southern river bank. The effect was further enhanced by a horizontal salmon ladder and part of the ritual consisted in releasing the salmon who jumped in large numbers through the reflected light.

Hugh Kearns and I carried out a survey by divining which verified the existence of lines of stones beneath the ground forming the salmon pools. This work was further verified by a team from the Irish Society of Diviners and a diagram showing the lines of stones can be seen in his book.[13]

From a five-sensory perspective water is regarded as a simple substance composed of hydrogen and oxygen. Water quality is considered from the viewpoint of chemistry and is concerned with removing contaminants and with disinfection by the use of chemicals such as chlorine. Yet there are subtle energy aspects to water that we do not yet fully understand. I have referred earlier to Victor Schauberger's imploded water and we have seen also that standing stones are associated with magnetic anomalies and underground streams.

Masaru Emoto has carried out research into the effects of our thoughts and actions on water. He takes samples of water and freezes them and then examines the ice crystals under a microscope. The results are startling. Water from a polluted river produced distorted crystals or no crystals at all but after the river had been cleaned up, beautiful hexagonal crystals were obtained. Similarly, when water was subjected to positive or negative emotions, beautiful crystals are produced from thoughts of love or happiness and distorted or even ugly crystals resulted from thoughts of war or hate. His book *The Secret Life of Water* contains fascinating photographs including a very distorted one of water from New York on September 11, 2001. [14]

The shaman was concerned with everyday practical aspects of life. Higher knowledge gained in trance states was applied to heal sickness, to advise on which crops to plant and to predict the weather. This was done in the vision quest, a search for knowledge beyond the ordinary senses. The initial trance state may have been induced by drumming, chanting or the taking of drugs but ultimately, whether induced by these methods or not, the shaman entered the Newgrange mound and cut off all sensory input by staying overnight alone in darkness or even being locked in by the rolling of a stone over the entrance.

It is well known that people hallucinate when deprived of sensory input. Hallucinations are often regarded as fantasies like dreams and of no value. But, as Stanislav Grof has shown, these experiences are universal and represent other states or realms from which information can be obtained. The Sacred Trust is a shamanic group in the UK affiliated to Michael Harner's Foundation for Shamanic Studies. It runs a course in which people are blindfolded and spend a weekend deprived of their normal visual input. For a time they may also be in silence. Katy Weitz attended one course for the *Observer* newspaper and described her experience.

> Suddenly I felt my whole body vibrating. I was transported to a forest where I was walking with my friend and then we were running…but now we were stags. Then owls. Then stars. This incredible vision continued to unfold for what felt like hours and though I would like to recount it all, it's far to long and makes me sound mad. But trust me, it really was fantastic.[15]

While such experiences may be "fantastic" they are not to be dismissed as fantasies and can have considerable value to the trained initiate. That is why such experiences and activities formed an important part of ancient rituals. The importance of ritual is in making the non-physical real. In subtle energy terms, thoughts are things and, as mentioned above, have a certain vibration or resonance. The projection of the thought form into the energy field is a precursor to its fulfilment. Worship and prayer are considered to be much more powerful when conducted in groups and in conducive places and atmospheres.

But what about other early rituals such as appeasement of gods and sacrifice? While I have proposed that Newgrange was inhabited by a partnership society we cannot assume that everyone was perfect and that disputes, disasters and illnesses were unknown. For a people who assign a personality to things that we regard as inanimate, we must also assume that a level of superstition existed at least among the less enlightened inhabitants. A superstitious people often see the need to appease angry gods. However, the finding of some cremated bones does not necessarily suggest sacrifice. It is more likely, knowing what we now know from modern shamans, that the bones were associated with rituals of contacting the spirits of the departed elders.

If we accept that the purpose of the Neolithic mounds was for the carrying out of shamanic practices for healing, vision quests and work through subtle energy transformation we can ask if the shamans tested the effectiveness of their mounds as they developed them. This testing might have been subjective but it is possible that when a new mound was built, a shaman tried it out to see how good it was at producing the correct environment for shamanic practice. Thus we can see how mounds might have been evaluated and suggestions made to try different forms of construction. We could imagine, given the skills of these people, that they might have sought advice in a trance state from higher beings on methods of construction or the use of subtle energies.

Looking at Knowth we could surmise that the seventeen satellite mounds were tested by the shamans as part of the development programme to find the optimum construction. Knowth could also have been a centre for shamanic training. The satellite mounds could have been used for training a number of initiates simultaneously and comparing their experiences. The two passages on opposite sides of Knowth could have been used for training two students who would have been out of contact with each other but physically close together. Their experiences could then have been compared. Alternatively, fully trained shamans may have worked in twos using the mound to see if they could gain more certainty in their insights if their experiences matched.

The Danish archaeologist Palle Eriksen in a paper called "The Great Mound of Newgrange"[16] suggests that Newgrange itself was developed in stages. He believes that the layers of turf grew naturally on top of each layer of stones and that this happened over a number of separate periods of development. This does not contradict the possibility that the layers were energetic in purpose. It is quite possible that the mound builders tried one layer and tested it for a while and then decided later to add others to increase the energy of the mound. Incidentally, Eriksen is also very critical of the white quartz wall. He says Newgrange has been "defaced by an exceptionally severe restoration".

From this chapter we can see that there is more to the use of Newgrange than just energy and healing. The issue of states of consciousness and the possibility of communication of thoughts and intentions through a universal energy field must be addressed if we are to proceed further.

Chapter Twelve

The Universal Energy Field

The process of healing which may have taken place at Newgrange may be considered to consist of two components – energy and intention. There is an intention to heal which produces a healing energy. The process of divination where the shaman looked for guidance may include an information component. Thus the practice of shamanism appears to access a universal field of information and energy. It is hard to accept this fact if you believe that we are all entirely separate from each other and that our brains do not receive any information except through our normal five senses. However, scientists have now proved that consciousness is not confined to the brain.[1] In an experiment rats were trained to run a maze and then parts of their brains were successively removed. Even though massive amounts of their brains were gone, they still retained the memory of how to find the way out of the maze. Scientists have suggested that memory is not just a question of accessing something in the memory banks of the brain but of using the brain as a receiver to connect to a universal memory bank.[2]

While the energies involved may be subtle and their existence difficult to prove, many examples of humans responding to a group energy field can be observed. Why do people riot? How is it that large groups of normally law-abiding citizens can go on the rampage? How is it that a powerful orator can whip up the feelings of a crowd to a frenzy? On a positive level we wonder how a talented musician can hold a hall of three thousand people spellbound. The 'spell', whether positive or negative, is the collective force that affects people and is caused by the interpenetration of people's auras. Spells are associated with magic and so it is not surprising that the concept of a human field of energy is not easily accepted by scientists.

Aura measurement, being considered unscientific, is consigned to the realm of paranormal research. But techniques for measuring auras have been developed and are being further improved. One such technique is called Kirlian Photography. Photographs of living objects such as leaves show discharges from their edges and when part of a leaf is cut off, the discharge from its aura still retains its original outline thus proving that it is related to something more than its physical shape.[3] It is well known that amputees experience phantom sensations from limbs that are no longer present. In another technique called Polycontrast Interference Photography, which I will refer to again later, photographs of subjects show different colours around parts of the body and these can be linked to the state of physical and mental health.

Rather than talking about auras, scientists like to deal in concepts such as "the extended mind". To investigate such phenomena PEAR, the Princeton Engineering Anomalies Research laboratory was set up by Robert Jahn, a professor of Engineering at Princeton. With private funding he carried out studies which showed clearly that humans could influence the operation of physical devices and processes simply by concentrating on them and forming intentions. Using various randomly occurring processes it was shown that consciousness creates order out of disorder by reducing the level of randomness of events such as the falling of objects and radioactive decay.

Further work carried out by Jahn and his colleagues led to the development of electronic random event generators (REGs) which enabled a wider range of experiments to be conducted. Normally such a generator would produce an equal amount if two opposite kinds of outputs called "highs" and "lows". A subject would sit in front of the machine and will it to produce more highs or more lows. Because the electronics allowed a large number of events to be produced, the ability to produce results which gave a larger sample size statistically was enhanced. The results showed that instead of getting a 50/50 outcome as expected by chance, the ratio of highs to lows was 51/49 in one series of tests and 54/46 in another. These results are not so impressive in the strength of the interaction between the individual and the machine but they are highly significant statistically, suggesting odds of a trillion to one that the results were not due to chance.[4] Thus we could say with confidence that the machines were influenced but not strongly influenced.

Having studied the REG trials, psychologist William Braud set out to see if the thoughts of one person could have an influence on another person. One group of volunteers was asked to direct calming thoughts at another group and were found to be almost as successful as when the group tried to calm themselves. Factors which helped the effect were a belief that it worked and a quiet mind. Braud had studied ancient Vedic texts which demonstrated how supernormal effects such as levitation could be produced by a quiet mind in the state of deep meditation. He suggested that this form of thought energy is in fact normal but masked by the constant state of noise and activity which surrounds us in the modern world.[5] As stated in Chapter 5, I have personally taken part in this form of meditation and can confirm that levitation effects can be produced and that I have experienced them myself.

The Global Consciousness Project,[6] a successor to PEAR, was set up at Princeton in an attempt to detect activities in the field of global consciousness. It has confirmed that we are all connected by an energy field which links us at a subtle level. In the Global Consciousness Project sixty five random number generators have been placed at locations around the world to act as detectors of consciousness in each area. Information from these generators is sent to Princeton and analysed by computer to see if there is any lack of randomness which would indicate a pattern associated with some major event in the world.

Most of us remember what we were doing on Tuesday 11th September 2001, the day the hijacked American airliners were deliberately flown into the twin towers of the World Trade Centre and the Pentagon. I had gone to the local library to look up some references and was just arriving back at the office in the early afternoon when I was told the news. It had happened about an hour earlier, (8.45am in New York); a portable TV was switched on in the office and I watched in horror as smoke poured out of both towers. I knew that they were designed to withstand an aircraft crash but I did not realise that the fire was so intense as to cause a progressive collapse within two hours. This traumatic event, witnessed by millions on TV as it happened, had a profound effect on human consciousness. Never before were so many people connected together by the intensity of their shared emotions. On 11th September the researchers in Princeton detected a significant pattern. Starting at the time of the first crash, the random numbers began to show statistically significant deviations from their normal expected pattern. This continued over a number of days.

The idea that individuals could be connected by a field of consciousness has not been accepted until recently. The Global Consciousness Project have the following statement on their website: http://noosphere.princeton.edu/

> Research on anomalies of consciousness shows that we may have direct communication links with each other, and that intentions can have effects in the world despite physical barriers and separations. Evidence compels us to accept correlations that we cannot yet explain. It appears that consciousness may sometimes produce something that resembles, at least metaphorically, a nonlocal field. The Global Consciousness Project (GCP) takes this possibility as a starting point for a speculation that such fields generated by individual consciousness would interact and combine, and ultimately have a global presence. Usually, because we are busy with individual lives, there is little to produce structure in the field, so it is random and not detectable. But occasionally there are global-scale events that bring great numbers of us to a common focus and an unusual coherence of thought and feeling. To study the effects of a possible global consciousness, we have created a world-spanning network of detectors sensitive to coherence and resonance in the mental domain.

The graph overleaf, reproduced by kind permission of Dr. Roger Nelson of the Global Consciousness Project, shows the results from analysing the output of the detectors starting ten minutes before the first crash and continuing for four hours after it. Normally the graph shows a line of random fluctuations centred about the zero line of normal expectation. The times of major events are indicated by the squares on the zero line. Here it can be seen that the output starts to deviate from normal at the time of the first event and continues to rise as more and more people began to hear about it. Statistical analysis showed that the probability was 35 to 1 that this was not just a random fluctuation, that it had some definite cause. As the anomaly continued over the next few days the statistical validity of the event improved. The upper line indicates a statistical significance of 0.05. which was passed just before 13.00 hours EST. The results show clearly that there is some form of connection between people at a very subtle level. The researchers described the results as mysterious. "We do not know if there is such a thing as a global consciousness", they said "but if there is, it was moved by the events of September 11, 2001".

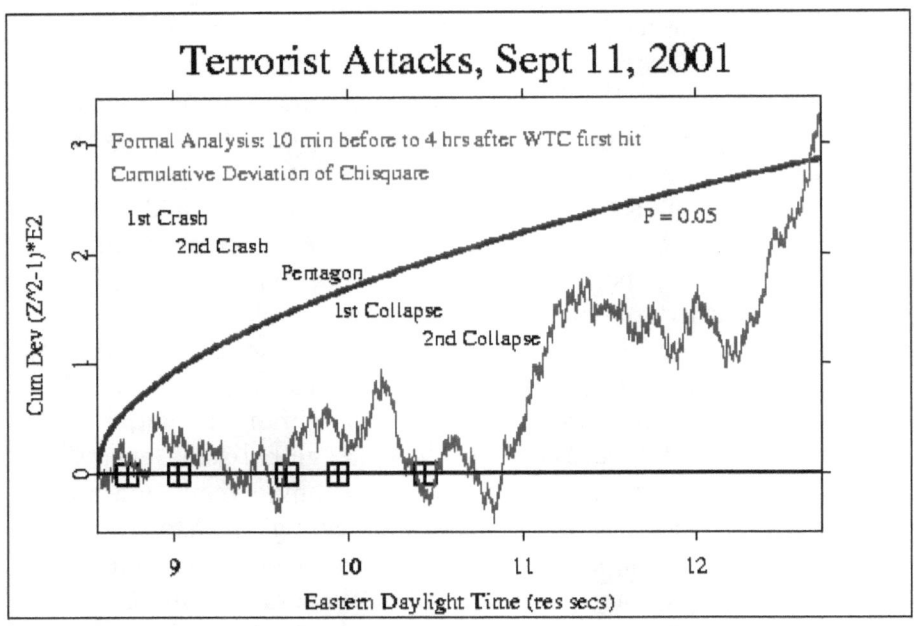

Terrorist Attacks, Sept 11, 2001

After September 11, a large number of meditators arrived at Maharishi University of Management (MUM) in Fairfield, Iowa to perform an advanced meditation for healing in society. The effect of the group meditation showed up in Princeton as a deflection from randomness in the energy field. The figure overleaf shows the data for Sept 26, the day the number peaked at 1,800 people. "This is the day of the maximum number of meditators, and the departure from expectation is steady and unusually strong, leading to a final result that has a chance likelihood of about one in 1000 (Chisquare 43190, df 42300, p = 0.0012) had it been an *a priori* prediction instead of an exploration." [7]

Research has shown that a person's energy field becomes stronger during meditation. Transcendental Meditation has been shown to be a settled state of awareness where the brain settles down to a state of enhanced alpha and theta rhythms. This state has been described as a unique state of

104

consciousness, a state of restful alertness.[8] The aura of a person meditating radiates a very calming or settled vibration which interacts with the auras of nearby people. Early studies on the field effects of meditation showed that people meditating in one place had an effect which could be detected in other meditators at a distance by measuring their brain waves with an electroencephalograph (EEG).

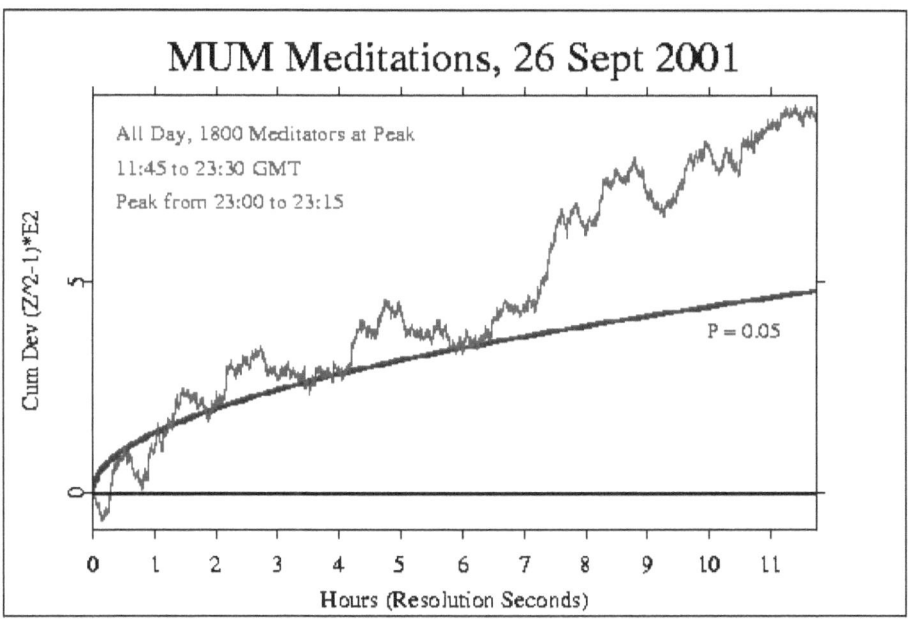

In 1979 a group of 2,500 advanced meditators assembled at Amherst, Massachusetts. During the several weeks that the group was meditating together, tests were carried out using three meditators in Fairfield, Iowa, 1,170 miles away. The purpose of the experiment was to try to detect increased synchronisation of brain waves as a result of the large group meditating. Each of the three meditators was placed in a sound-proof room and neither they, nor the laboratory technicians, were aware of the times when the group in Amherst was meditating. The study showed conclusively that when the group in Amherst meditated, there was increased synchronisation between the brainwaves of the three meditators in Iowa.[9]

Another graph from the Global Consciousness Project is reproduced overleaf. This reflects the wave of positive emotion following the election of Barack Obama as president of the United Sates on November 4th 2008.

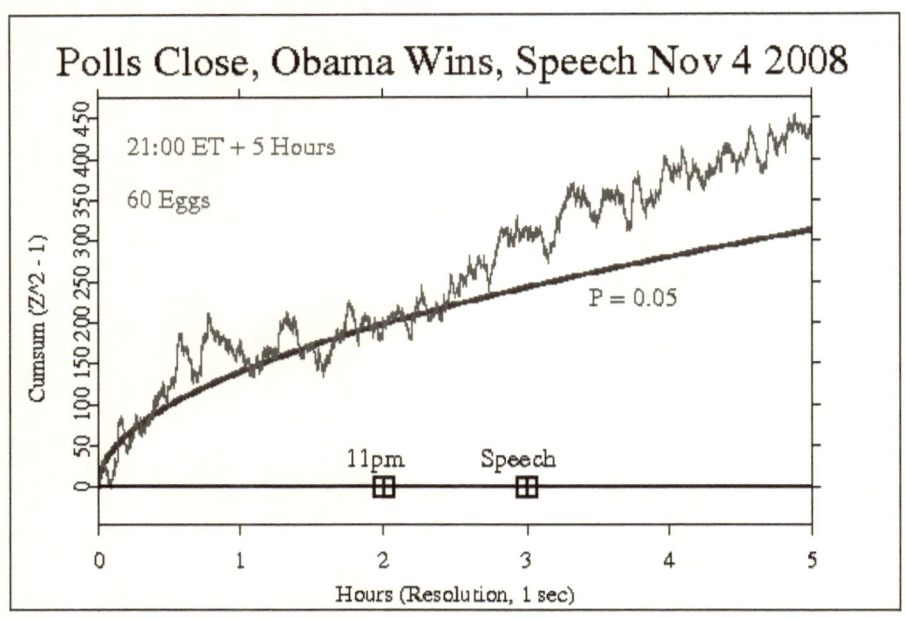

There have been numerous other cases reported where action at a distance has taken place without direct physical contact between the people involved. These give a clear indication of the existence of a human energy field and show its potential for healing the body as well as healing society. When I was at the Prophets Conference in New York one of the speakers, a scientist named Gregg Braden, showed a video of a healing using Qi Gong which took place in May 1995 at China's largest medicineless hospital, the Huaxia Zhineng Qi Gong Centre. A lady with a bladder tumour was being constantly monitored by ultrasound. The tumour could clearly be seen on the screen. Four practitioners of Qi Gong started chanting loudly and insistently and the tumour started to get smaller. After three minutes it was gone completely. The treatment was carried out by placing the hands over, but not touching, the patient.[10] Braden assured us that the video was genuine and that he had thoroughly checked its veracity.

Two separate studies have shown that we can send thoughts to consciously influence our DNA and the DNA of others.[11] DNA is short for deoxyribonucleic acid and is a structure that contains the information for the development and functioning of all living organisms. It consists of two strands of substances called nucleotides twisted around each other in the form of a helix or type of spiral in three dimensions.. In one experiment individuals focused their intention to either wind or unwind the two strands which make up the DNA double helix. The DNA was in the form of human placental DNA, the most pure form available. The winding or unwinding was measured by UV absorbtion spectroscopy. Samples treated with the conscious intention to unwind DNA caused a larger increase in absorption than those not treated. These effects, which are statistically significant, indicate that DNA is sensitive to the informational content of consciousness. It seems that negative emotions cause DNA to wind into tighter bundles and positive emotions cause DNA to relax. This may have implications for research on healing. It suggests that the genetic code is not fixed and may be changed by the quality of human feeling and emotion.

Many people have seen the film *What the Bleep do we Know?* In this film the physicist Dr. John Hagelin describes a demonstration project where 4,000 meditators gathered together in Washington in 1993 and in a scientific study proved that by meditating together they had produced a 23% reduction in crime. This study was published in the journal *Social Indicators Research*.[12] He says that they knew it would be successful because of 48 previous studies which had demonstrated the effect of large groups of meditators in producing a calming effect on society. Many of these studies had been published in respected academic journals. One study published in the *Journal of Conflict Resolution* in 1988 showed how a reduction in conflict in the Lebanon was brought about.[13]

If these studies produced results which have such profound implications for healing society why were they ignored? The reason is that they did not form part of people's reality at that time. *What the Bleep do we Know?* explains that we create our reality out of our consciousness. If something is not part of our experience, we cannot understand it. The brain receives over 400,000 pieces of information but only lets through 2,000. These studies were ignored because journalists and people of influence, especially academics, had great difficulty accepting the fact that there could be a connection between meditating and reducing stress in society. As I said earlier concerning the lessons of Newgrange, it is not enough to prove that something is a scientific fact. You must also explain the theory behind it in a way that is compatible with the level of knowledge prevailing at that time. Another reason why the studies were ignored was the issue of culture. Journalists in particular found it difficult to accept something coming from India and associated with a guru. Lynne McTaggart refers to this in *The Field*, an important book reviewing the latest research into subtle energies:

> Although the TM organisation has been ridiculed, largely because of the promotion of the Maharishi's own personal interests, the sheer weight of data is compelling. Many of the studies have been published in impressive journals such as the *Journal of Conflict Resolution*, the *Journal of Mind and Behavior*, and *Social Indicators Research*, which means that they would have had to meet stringent reviewing procedures.[14]

Although the results of the Washington project were ignored it seems that they are now no longer being treated with scepticism. McTaggart states that scientists and others are beginning to recognise the existence of a universal energy field. Physicists are beginning to be able to measure it, as we have seen, and some recognise it as a new kind of field - a torsion field. Some scientists think that it could also be a field of human consciousness.

In the previous chapter I argued that the reality experienced in shamanic states is outside our normal conscious experience. In this chapter I have shown evidence that the human mind is able to transmit energy and intentions across vast distances and is much more powerful than we could ever imagine. I hope to apply these conclusions to demonstrate that the reality experienced in Neolithic Ireland may have been totally different to anything we can imagine and that we should not try to analyse it by comparing it with our present reality. In particular we have to be careful when applying the current scientific method of analysis to this kind of subject. For example, the scientific community is totally dismissive of astrology but in the next chapter I hope to show that the people of Newgrange were aware of subtle planetary energies and may also have received information from other ancient civilisations.

Chapter Thirteen

As Above, So Below

In the old world, the solstice was considered a time when the veil between this world and the 'otherworld' became very thin. This traditional belief is seen in the history and myths of many places but particularly of Egypt and Peru.[1] Ancient sites were regarded as portals to the otherworld and the time when the 'sun stood still' (the meaning of solstice) was the time when those with the most highly developed psychic abilities were able to penetrate the veil and obtain knowledge from the higher dimensional otherworld.

Astrologers are fond of using the old saying "As above so below" to explain how the planets could affect our daily lives. There is increasing evidence that ancient societies possessed advanced astronomical skills and used them in their quest for higher wisdom. In their book *The Orion Mystery* [2] Robert Bauval and Adrian Gilbert show how the layout of the Pyramids of Giza in Egypt accurately mirrors the stars in the constellation Orion.

It is well known that the Great Pyramid contains shafts from the 'burial chambers' which are aligned with certain stars at certain times. For example, one of the shafts in the King's Chamber was aligned with the position of the star Zeta Orionis, the lower and brighter of the three stars in Orion's Belt at about 2,500 B.C., the accepted date of construction of the pyramid. It is not proven that this was a burial chamber as it was found to be empty when first discovered in modern times.[3] In fact the commonly held belief that Egyptian mummies were found in pyramids must be examined carefully. No pyramid chamber has ever been discovered which contained a mummy which was an original burial, but there are cases of pyramids being used as burial chambers long after their original construction.[4]

If the pyramids were not originally designed as burial chambers what was their purpose? There has been much speculation on this subject and some considerable research by people who were not archaeologists. The theories range from energy generators to pumps to astronomical

109

observatories. Whatever their purpose, it is clear that the pyramids were constructed by a highly advanced society with an amazing knowledge of astronomy and mathematics, far in advance of that which was thought to exist at the time.

The King's Chamber of the Great Pyramid is reported to resonate to sounds at a frequency of 438 Hz. Christopher Dunn believes that the chamber was tuned as a massive acoustic resonator. The granite used in its construction contains a high proportion of quartz crystals. Quartz has the interesting property of being piezo-electric. This means that if it is vibrated it will give out an electric current. Similarly, if fed an electric current, a quartz crystal will vibrate. This effect is used in small loudspeakers for computers and mobile phones.

Many strange effects have been observed in the pyramids. Dunn describes how a wine bottle became charged with electricity and gave off sparks. Cats which get trapped in the pyramids and die do not decay. Their bodies become dehydrated. Model pyramids have been made and used for duplicating some of these effects and are claimed to have healing properties.[5] The acoustic properties of the King's Chamber are very strange. Many people have lain down and chanted in the large rectangular box which the archaeologists have called a sarcophagus although it is quite different from those found in burial chambers. Mike Pettigrew of the Institute for Afterlife Studies gave me an audio tape where he describes his experience in the sarcophagus.

> As I lay down and started toning, the sound of my voice seemed to bounce back and forth and up and down in the sarcophagus. The sound of my voice with every reflection became louder and louder until my entire body was vibrating in resonance.[6]

The pyramids of Egypt are only one of many examples of astronomical alignment, massive construction, or acoustic resonance effects. Stonehenge is another and so of course is Newgrange. At the beginning of this chapter I mentioned that astrologers are fond of using the old saying "As above so below" to explain how the planets could affect our daily lives and that there is increasing evidence that ancient societies possessed advanced astronomical skills and used them in their quest for higher wisdom. People are right to be sceptical about astrology where it is presented as horoscopes in newspapers. These assume that the population of the world can simply be divided into twelve groups based on the twelve signs of the zodiac and that predictions for such large groups could have validity for each individual. However, I have found certain types of astrology, particularly the Indian version, to be remarkably scientific in its approach and accurate in its predictions. If you are still sceptical, I ask you to postpone judgement for the moment while we delve into the possible role of astrology in the lives of the builders of Newgrange.

Let us use our imaginations to create the reality experienced at Newgrange in 3,200 B.C. These people lived in a warmer climate and would have been used to looking at the stars at night. If you have ever camped in the countryside without any electricity you will immediately remember how familiar the night sky becomes after only a few days. Ancient peoples all over the world saw patterns in the night sky and named the constellations. A remarkable fact is that the same twelve signs of the zodiac were used by many ancient peoples. The earliest references to astrology are in the Vedas. These writings represent an oral tradition going back into the mists of time, impossible to date with any accuracy but certainly much older than Newgrange. Although the people of Newgrange did not leave a written legacy we cannot assume that they had no education or science. In their book *Light on Life: an introduction to the astrology of India*, Hart Defouw and Robert Svoboda state:

> The written word was an afterthought, not an ideal for many civilisations. In India, traditional authorities assert that the custodians of sacred knowledge recorded it in written form only when its transmission became endangered because fit disciples could no longer be found… We belong to a culture which worships the visual and this makes it difficult for us to conceive that knowledge was once transmitted orally from generation to generation with accuracy and that writing was once regarded as a crutch of little intrinsic value. It is possible that not only were our ancestors' minds comparable to our own, they may even have been superior to ours in many ways.[7]

There are six sub-branches in the Vedas and these are called the "Vedangas" or Limbs of the Vedas. One of these is called "Jyotish" which is astrology and astronomy. The name derives from the Sanskrit word *jyoti* meaning light. While a superficial meaning is the study of the lights in the sky, it is clear that early astrology was based on observation and experience of connections between the movements of earthly bodies and events on Earth. Thus it was said that Jyotish sheds light on the past, the present, and the future.

The basic principles of astrology and astronomy seem to have passed from the Vedic civilisation through the Persians and Babylonians to the Greeks and thence to the Western world.[8] It is interesting to note that the two subjects were one until the seventeenth century. At this stage the more scientifically minded insisted in relying only on that which could be observed and proved, and the experiential and intuitive aspects were abandoned.

While it cannot be proven that astrology was practised at Newgrange it is clear that astronomy was and we know that the two subjects were one until relatively recent times. The reality at that time bore witness to the connectedness between all things through subtle energy fields, and we must assume that astrology was understood and practised even if no written records exist. It may not have recognised the same signs of the zodiac, but it recognized the energy values associated with the planets and constellations and their effects on human life. The mound at Newgrange may have been part of an astronomical observatory, but the main work of carrying out astronomical observations and constructing calendars could have been done using the standing stones which surround the mound and the wooden henge or circular monument beside it.

The primary purpose of astrology was the determination of auspicious times for certain events to take place.[9] These ranged from rituals and sacrifices to planting crops, celebrating weddings and starting projects like constructing buildings. Astrology is still widely used for these purposes in India. It is considered that each planet and the other heavenly bodies such as the sun and the moon have an energy which has certain qualities. We know that the sun and the moon exert gravitational and electromagnetic forces on the earth. Effects such as disruption of communications by sunspots and the influence of the moon on the tides are well known. The electromagnetic fields of the sun and earth are toroidal or doughnut-shaped and intersect each other. Many people accept the influence of the moon on their moods and notice how their mood changes with the arrival of the full moon. I mentioned in Chapter 3 how Laurence Edwards showed that plants grow according to astronomical cycles.

In a monograph entitled "Living the Field: Earth Energies", Lynne McTaggart has reviewed the scientific research on the effects of the sun, moon and planets on humans and animals.[10] Many of the earlier research studies mentioned gave conflicting results but it can be seen that in these cases the research design was too simple. For example, researchers looked for direct correlations with the full moon or new moon when it is clear from an energy perspective that the effect of the moon is not just experienced at the extremes but builds up gradually as the moon waxes and wanes. More convincing results have been obtained by researchers who looked at the positions of planets at the time of birth of peak performers. A significant Mars effect was shown by the French psychologist Michel Gaugelin. He found that many sports personalities were born within two hours of Mars rising or two hours after it crosses the mid-heaven. McTaggart lists many attempts to discredit this research and some which supported it. Gaugelins' work was reviewed by Suitbert Ertel, a German psychologist and with improved research design, the results were confirmed. In many cases better correlations were obtained and planetary effects were found for doctors and some other professions.

It is clear that the effects of the planets on our lives can no longer be dismissed as fantasy. Interactions between the electromagnetic and gravitational fields of the heavenly bodies have been shown to produce effects such as changes in the shape of the earth and the tilt of its axis. The earth has locked into gravitational resonances with the other planets and the cumulative effects of these resonances combined with electromagnetic fields over time have been shown to have profound effects on climate and on living organisms. Each of the planets produces a vibration of a certain frequency, and studies have shown that the interactions of these frequencies and higher harmonics can affect the intelligence of children and their personalities. Other studies have shown correlations with accidents and earthquakes.[11]

In terms of subtle energies, effects and much finer qualities in the energy of each of the heavenly bodies have been noted and the knowledge refined by astrologers over many years. Some planets are seen as malefic and some as benific. In India certain planetary placements are well known, such as *kuja dosha* which refers to Mars and its effect on a person's marital happiness. An astrological chart is divided into twelve sections called houses corresponding to the twelve signs of the zodiac. Depending on your time of birth the planets are seen to occupy particular houses. If your chart has Mars in the first or seventh house, it will be difficult to marry because this placement is universally accepted as being unfavourable to marriage. Such a person is described as "mangalic" as *Mangala* is the Sanskrit for Mars and many Indian astrologers or Jyotishis, offer remedies and advice for such situations. Skilled practitioners of Jyotish learn off vast numbers of planetary combinations and are able to give very detailed readings based on a person's place and time of birth.

Although my study of Jyotish is at a very elementary level compared with the Indian Jyotishis, I have done a number of charts for friends and relatives and have surprised both them and myself by their accuracy in predicting some future events. From the viewpoint of physics, time is merely a mathematical concept. For example, if we know the magnitude and direction of certain forces we can calculate backwards to see where they started or forwards to see their final effect. Using the sophisticated computer programmes now available for studying astrology we can roll time forwards and backwards and begin to see time as a continuum rather than as a fixed point in which we are confined. Time is really a method of expressing change. If time stood still nothing would change and if nothing changed, time would stand still. Beyond space-time is just one eternal "Now", and the part of us that can transcend space and time is thus able to access information about the past and the future, and is also apparently able to obtain information across vast distances.

I believe that the elders who inhabited Newgrange had an extensive and intuitive knowledge of astrology which they used in conjunction with their vision quests to help and advise their people. This knowledge made its earliest appearance in India but I believe that it was transferred intuitively to other ancient peoples. Evidence of the intuitive transfer of knowledge through energy fields comes from the work of biologist Rupert Sheldrake.

It has been observed in nature that once a certain behaviour has been learned it becomes much easier for others of the same species to learn it, regardless of distance. One example of this comes from the behaviour of birds. In Britain in the early part of the last century it was noticed that tits in one area had discovered how to tear the caps off milk bottles and drink the top of the milk. Although tits never travel more than a few miles, it was soon found that tits in isolated pockets all over Britain were doing the same thing. The habit then spread to Sweden, Denmark and Holland and died out during the Second World War when doorstep milk deliveries temporarily stopped. After the war it started up again, even though all the earlier birds had by now died.

Sheldrake's seminal book *A New Science of Life: The Hypothesis of Formative Causation* was published in 1981.[12] It was described by *Nature* magazine as "the best candidate for burning there has been for many years" and yet became a best-seller. Resisting the training which attempted to teach him that Nature is dead, Sheldrake has proposed that self organising systems including molecules, crystals, cells, tissues, organisms and societies of organisms are organised by 'morphogenetic fields', energy fields whose effect can be felt through space and time. These fields provide the basis for the principle of formative causation whereby a behaviour learnt by one individual of a species is made available to all others of that species independently of physical contact.

The name morphogenetic comes from the Greek *morphe* meaning form and *genesis* meaning coming into being. So morphogenetic fields have the power to cause organisms to follow a certain pattern of development. They are a sub-set of a wider range of fields which Sheldrake calls morphic fields through which organisms and even inanimate matter apparently communicate.

By the process of morphic resonance a field is created whereby 'like influences like' through space and time. "Morphic resonance", he states, "does not fall off with distance. It does not involve a transfer of energy, but of information".[13] He claims that it is well known that some new organic compounds are difficult to crystallise, yet, as time goes on, they tend to form more readily all over the world. He relates how chemists, when asked to explain this phenomenon, have resorted to the most unlikely theories. Some have suggested that fragments of crystals must have been transferred from laboratory to laboratory on the beards or clothes of scientists.

According to Sheldrake morphic fields are responsible for the behaviour of flocks of birds which turn together as if they were one super-organism, the wave of information passing instantly from one side of the flock to the other. They are also responsible for the instinctive behaviour of animals. These fields are of a type that has not been measured by physics and are evolutionary in nature. As one species learns to do something, others of the same species learn it more quickly.

As well as applying to animals, these fields also apply to human societies and to inanimate objects such as crystals, so Sheldrake argues that the laws of nature are more like habits maintained by morphic resonance. Instead of the laws of nature being seen as external to the world, they are inherent in it, reflecting the collective memory of each species. The more often a pattern of behaviour is repeated, the more likely it will be repeated again.

Sheldrake sees time as having a morphic resonance which links the present and the past. For example he describes festivals and rituals as being intensely conservative. By repeating the exact formula as conducted in the past, "the present participants are linked to all those who have gone before – to the ancestors".[14] He has also extended his ideas to cover places. The concept of a field, he says, is grounded in the idea of place. Morphic resonances of a disturbing quality will be experienced in places where murders have been committed whereas a more uplifting quality will be experienced in sacred places.

> All over the world certain places are regarded as sacred. Such holy places may be natural sites such as springs, mountains, and groves of trees on hilltops, or they may be places where standing stones, stone circles, tombs, shrines, temples, churches or other buildings have been erected...... And their orientation often relates them to significant natural features such as the point on the horizon where the sun rises at midsummer....Such sites are

venerated because of what happened there in the past. They are places where sacred experiences or revelations have occurred; where heroes and saints were born, lived or died; or where their remains are preserved. What happened in the past can in some sense become present there again and thus they can act as doorways to realms of experience that transcend the ordinary limitations of space and time.[15]

In a book entitled *Dialogues with Scientists and Sages; The Search for Unity*, Sheldrake, David Bohm and major spiritual figures such as the Dalai Lama engage in discussions with the philosopher Renée Weber.[16] In a discussion with Sheldrake she asks him how he would account for the power of a rare individual such as Buddha or Christ since he states that the strength and dominance of a morphogenetic field is related to the number of organisms who have built it up. Sheldrake replies that there is also an intensity factor, a qualitative as well as a quantitative element. He says that the "consistency and intensity of a pattern of thought or an intention strongly held creates the power in the field". Weber then refers to her earlier discussion with Bohm.

> When I discussed this with David Bohm, he felt that the power of a Buddha-field, disproportionate to its numbers, came from the level of wholeness deep within the implicate order from where the Buddha operated. At that level only compassion and order are possible not hatred and disorder.[17]

Thus we are all connected together at some level both to each other and to the cosmos. This surely is one of the attributes of a partnership society. In such a society people are aware of their connectedness. Today we seem to have lost this awareness but maybe the people of Newgrange were consciously aware of it at all times. In fact their reality was so different that their consciousness had not created some of the things we take for granted today. With enough of everything to go around, there was no need to create the concept of individual wealth so there was no need to hoard. There was no such thing as putting yourself first because the individual and society were one. It is hard for us to imagine living like this but we should try to do so if we are to really understand what their life was like.

From this understanding we may be able to learn how to recreate a partnership society and how to use its energies to heal ourselves and the planet.

Chapter Fourteen

The Energy Field is Damaged

In the previous chapters I have described various manifestations and explanations of a field of subtle energy which is everywhere about us and which, is generally known as the Universal Energy Field (UEF). This is a field of energy which is not normally detectable by conventional scientific instruments and its effects are subtle and only vaguely understood. In the early days of the discovery of electricity, people were able to produce static electricity by rubbing fur on amber. They could produce some sparks but did not understand how electricity worked and could not control it. This is where we are in our understanding and use of the UEF.

Static electricity was so called because it built up as a stationary charge and was not readily usable. Since the days of static electricity we have learned how to make electricity flow and how to use the energy it contains. It now forms an essential part of modern living. The same thing will happen with the UEF. As time goes by we will learn how it works and how to use it for our benefit. It could provide us with the ability to free ourselves from illness, to control our weather and to access unlimited amounts of energy. It could also deliver a peaceful society like that which existed at Newgrange.

Let us summarise what we know about this field and see if we can learn something about how we might benefit from it. We know that it is related to electromagnetic fields and can be obtained by screening out the electromagnetic fields and seeing what is left. We know that it is related to the property of spin and is associated with spinning energy. It may be generated when matter is spun very fast or accelerated in a vortex. We know that it may be associated with magnets and that spinning magnets and gyroscopes can generate anti-gravity effects. It seems to be communicated from one object or person to another by the principle of resonance, of things vibrating at the same frequency, and it is greatly strengthened by this quality of waves being in phase with each other, a phenomenon which is called coherence. Once it has been communicated, it can imprint any object with a record of the field.

Some physicists call it a scalar field, meaning that it has magnitude but no direction. Others call it a torsion field indicating that it is spiral or helical in form. There is much evidence of such a field in Nature in the growth of organisms, the spiralling of galaxies and the destructive power of hurricanes. Three dimensional spiral forms such as the torus or doughnut shape have been shown to possess remarkable stability in the form of smoke rings. In fact scientists have shown that the magnetic fields of the earth and the sun are toroidal, as is the magnetic field of the human heart, which extends up to fifteen feet from the body.[1] Since there is a relationship between magnetic fields and the UEF, some scientists have postulated that the detection of the magnetic field of the heart is an indicator of the presence of the UEF which is also toroidal or spiral in form. This form of the UEF is called the Human Energy Field or HEF and is detectable by sensitive instruments. But it is also delicate and may be affected by outside influences since torsion fields can imprint organisms. The presence of man-made electromagnetic fields may produce torsion fields and these may affect us at very subtle levels and in ways which we do not yet understand and which may be harmful.

In 3200 B.C. electricity, as we know it, did not exist and the strongest electromagnetic fields affecting the inhabitants of Newgrange were those produced by the earth and the sun. Recent research has shown that we are now exposed to electromagnetic fields 100 million times stronger than those that come from the sun. In fact it is claimed that we are suffering seriously from electromagnetic pollution, that the earth's energy field has been swamped by man-made radiation which is claimed to be responsible for much of our present problems.

Robert O. Becker showed how all living things can be affected by electric fields which are much weaker than the level of exposure normally regarded as safe. He graduated from New York University's College of Medicine in 1948 and became an orthopaedic surgeon. His pioneering research in the field of regeneration of bone and muscle after injuries showed the importance of weak electrical currents in stimulation of regrowth and healing. In 1985 in his remarkable book, *The Body Electric*,[2] Becker described his early research and warned the public about the growing threat of electromagnetic pollution in the environment. All through his career he was the victim of political decisions that restricted his research funds and many of his ideas are still not recognised by the medical profession. In latter years he came up against powerful government and commercial interests who were concerned that the public would object to the expansion of the electricity and telecommunications networks.

Becker pioneered many healing techniques using weak electric currents. He regenerated limbs in salamanders and frogs and showed how electric currents could be used to kill bacteria and halve the healing time for broken bones. In his book he quotes how by accident it was found that if a child younger than eleven years of age loses the top joint of a finger it will regrow perfectly, including a new bone, if the skin is not sewn up. Yet his techniques remain largely unrecognised. Broken bones are still routinely put in plaster without the use of electrical stimulation and surgeons still prefer to use complex microsurgery with limited success.

In 1990 Becker produced a book which concentrated on the dangers of electromagnetic fields. This book, *Cross Currents, The Perils of Electropollution, The Promise of Electromedicine*[3] identified how our bodies and immune systems are being adversely affected by man-made electromagnetic fields from power lines and the wide range of electrical equipment which we use every day. Becker claims that radiation, once considered safe, is now correlated with increases in cancer, birth defects, depression, learning disabilities, and illnesses such as chronic fatigue syndrome and Alzheimer's disease.

In an interview in 2000 on the subject of mobile phone emissions Becker said:

> There are definitive effects of low strength oscillating electromagnetic fields on brain function. Now, we look around at the present time and I have lived through roughly half of this period of increasing use of electromagnetic energies. We're looking at an entirely different behavioral aspect of the population at the present time. We certainly have a far different social attitude at the present time among the majority of the population...If you look at the proliferation of what is called Attention Deficit Disorder (ADD) - that wasn't even here when I was young. That was not a diagnosis. It never existed.[4]

Becker was particularly concerned about the effect of electromagnetic fields on growing children and was responding to a recent research report which showed that radiation from mobile phones had been shown to cause damage to proteins in worms. The report said that children in particular, were considered to be at risk, because their nervous systems are still developing and because the smaller size of a child's skull allows greater absorption into the brain tissue of the low level microwaves emitted by mobile phones.

Other groups and individuals have also expressed concern. The Childhood Cancer Research Group at Oxford have reported on a seven year study which suggests that children living near high voltage power lines have an increased risk of contracting leukaemia. The group studied 9,700 children with leukaemia and concluded, as reported in the *British Medical Journal,* that living within 200 metres of a power line is linked to a 70 per cent increase in risk, and living within 600 metres is linked to a 23 per cent increase in risk. The scientists were reported to be puzzled by the fact that the effect extended out as far as 600 metres.[5]

Lawrence Edwards whom I mentioned in previous chapters was also concerned with the effects of energy fields on growth and I have described his work on analysing the spiral growth patterns of plants using projective geometry. He found that plants grow according to unseen energy fields related to the positions of the sun, moon and planets. He performed many measurements of bud and tree growth and related the growth patterns to astronomical cycles. In the course of this work he observed that a particular tree beside a high voltage sub-station had distorted growth, and that its growth patterns, unlike those of the other trees, were unrelated to astronomical influences.[6]

Wilhelm Reich identified a destructive form of energy which he called DOR (Deadly Orgone Radiation) This energy had positive ions, signified a lack or stagnation of orgone energy and was created by TVs, microwaves, and other electromagnetic sources. It is well known that positive ions abound in so-called "sick buildings" and that negative ionisers have been found to alleviate the symptoms such as headache and fatigue complained of by the occupants.

David R. Cowan has spent over thirty years researching and healing the effects of electromagnetic pollution and unhealthy earth energies. In his book *Safe as Houses* [7] he echoes the comments of Dr. Becker:

> It seems very much that the pattern of disease has markedly shifted in this century. The most threatening diseases up to the 1950s were diphtheria, tuberculosis, influenza, polio, heart disease and some forms of cancer. Since then there has been a great increase in the immune deficiency diseases such as allergies, asthma, Chronic Fatigue Syndrome (M.E.), AIDS, arthritis, and cancers linked to the immune system like leukaemia, lymphatic, liver and intestinal.

As well as being affected by electromagnetic radiation we are also being subjected to other forms of energy fields which may be injurious to our health. These effects go under the general name of geopathic stress. As Cowan says, "Earth energies and leylines have a sinister side. When a leyline passes through decaying matter, like a burial-ground, dirty river or canal, the energy changes to a "black stream"... Apart from this, underground streams, geological fissures and fault lines, coal mines, quarries, electric power stations and sub-stations, microwave towers, repeater television aerials, nuclear submarines and missile sites, etc., all alter the natural energies around them". He goes on to give numerous examples of geopathic stress and shows how, as a dowser and geopathic stress consultant, he has remedied the energies in many people's houses and improved their health.

These theories are now supported by scientists who have developed the new science of geopathology. It is well known that certain areas are affected by radon, a radioactive gas which leaks from under the ground and can concentrate under buildings. These areas are associated with increased incidence of cancers. Also well documented is the effect of positive ions. These are atoms which have lost electrons and are responsible for the "sick building syndrome". They are found in modern office buildings and homes constructed using materials such as plastics and metals. The harmful effects of positive ions and corresponding beneficial effects of negative ions has been shown in over 75 papers by Dr. Albert Krueger a bacteriologist at the University of California.

Positive ions cause fatigue, depression and irritability. Dr. Roger Coghill of Coghill Research Laboratories explains that ions collect at sharp point above the ground such as the top of buildings:

> Tall buildings are subject to the same ion flow, and since the earth's surface is mildly negatively charged there should be slightly more negative ions at building tops. Mountains also contain more negative ions, and their beneficial effects makes them a holiday favourite, second only to the seaside, where "neg-ions" are also generated from the tumult of wave upon shore. Instinctively we seem to know that such places are good for us.

Less well known is the effect of underground streams. Coghill reports a number of cases where "clusters of myalgic encephalomyelitis (ME) nearly always occur first in dormitory residences near such subterranean aquifers". He cites cases such as the 1955 outbreak at the residential school for trainee nurses at the Royal Free Hospital, London, This building "was only a few metres from Fleet Road, a steepish hill down which runs the now conduited Fleet river on its way from Hampstead ponds to the Thames". He describes similar outbreaks at the Middlesex Hospital (1952) and the Children's Hospital, Great Ormond Street (1970). All of these locations were close to important underground streams or rivers which had been covered over, and are no longer used as dormitories.[8]

It appears that many buildings have been erected in places affected by harmful energies. The importance of building in sympathy with the energy of a certain place is recognised in many ancient cultures. In India they use the ancient knowledge from the Vedas to ensure that houses are built in a way that does not attract harmful energies. The system is called Sthapatya Veda and comes from the Sanskrit word *Stha* meaning 'to stand'. It deals primarily with the subject of 'Vastu', or the correct placement of a house and its rooms. This is considered essential for harmonious and healthy living. For example, the entrance to a house should face East, and the various rooms should occupy parts of the house most suited to the energy of the sun. Marcus Schmieke, a German physicist who has specialised in this subject, has studied this subject with Vedic experts in India and gives useful guidance in his book *Vastu*.[9] In general, Sthapatya Veda recommends that cooking should be done in the

Southeast corner and the main meal eaten at midday, the time of maximum digestive fire. Thus the dining room should be in the South. Bedrooms should face Southwest or Northwest and the living room should be in the West. One should have a meditation room in the Northeast.

The direction that the entrance of a house faces is considered to be very important because it affects the quality of the energy entering the house. If one cannot get a suitable house with an entrance facing East, then North is next best since it is considered to bring prosperity. The other six out of eight directions are regarded as less favourable. For example, a house with a South-facing entrance is considered to be the worst orientation since it is affected by the harsh quality of the energy of the midday sun.

Feng Shui, pronounced "Fung Shoy", is a Chinese system for creating a healthier living and working environment. It is claimed to be over 7,000 years old and has evolved gradually over the years, possibly borrowing some of its principles from Sthapatya Veda. It emphasises the correct flow of 'chi' or subtle energy inside and outside a building. Like Sthapatya Veda, it has a lot to say about the correct placement of a building but is not so concerned with the points of the compass. So, according to Feng Shui, you could live happily in a South-facing house provided it is correctly positioned in the landscape.

Feng Shui means wind and water, referring to the two fundamental elements, air and water, that we need to survive. Without breathing and without drinking water we will not last very long. Buildings are considered well placed, for example, if they have water in front and a hill behind. This seems to fit with defensive requirements and may not date from the principles of a partnership society. However, it seems to work and very few buildings in China are constructed without taking it into account.

In *The Feng Shui Handbook*, Master Lam Kam Chuen lays out the principles of Feng Shui in a clear and concise manner, both for the placement of a house and for the flow of chi in the rooms. He gives examples of buildings that are well placed and a few that are badly placed. The United Nations building in New York is pictured showing how it occupies an undesirable location, alone and unsupported by other buildings, and with water at its back. He leaves it to us to draw our own conclusions about how successful it is as a workplace.[10] Interestingly, a website on New York architecture has revealed the following quote:

> The United Nations is seeking ways to make extensive repairs and updates to the building that is plagued by asbestos, lead paint and concrete falling off, as well as inefficient ventilation and windows and a lack of fire sprinklers. Renovation could cost $1 billion and possibly require the construction of a new office tower to the south of 42nd Street to house the Secretariat's activities meanwhile.

It is only in the last one hundred years or so that mankind discovered how to produce powerful electromagnetic fields. Previous to that, and in the time of the builders of Newgrange, no such energy fields existed. We have to start asking ourselves if we really need to be surrounded by the present levels of electromagnetic radiation.

Electrical hypersensitivity is a condition that is now recognised in some countries, notably Sweden, which is leading the way in reducing exposure to unwanted radiation.[11] Some people seem to be more sensitive to electromagnetic fields than others and this may be a source of Chronic Fatigue Syndrome. The main symptoms reported by sufferers and which are connected to electromagnetic fields are: burning or tingling on the face, dryness of the eyes and mouth, heart palpitations, fatigue, and disturbed sleep. Because only some people are affected at a level where they have to seek remedies, it is difficult to convince governments that everyone is being affected but at lower levels, the effects of which may not manifest themselves until many years have passed and millions of people have suffered long term damage to their health.

With the increasing proliferation of wireless technology, we are being surrounded daily by even more electromagnetic fields which are all-pervasive. Present guidelines on exposure to electromagnetic radiation consider only the heating effects of such energies and it would be wise to conduct some serious research looking at such effects on organisms at a cellular level. It is clear that these electric fields distort the subtle energy fields which govern the growth of organisms. They may have other effects, perhaps even on fully grown organisms. The problem is that we

126

do not know how to measure these effects and yet the amount of electrical energy around us is increasing all the time, as if there was no limit to the amount we can absorb without being affected in some way.

It is just like the situation which applies to the dumping of waste. Companies and individuals considered that they could burn or bury their waste and that the environment would not be affected by it. It was only when the quantity of waste increased that it became clear that there was a need to set limits so as to avoid irreversible environmental damage.

In the next chapters we will see what we can learn by combining the knowledge of the ancients with more modern ideas. Maybe this knowledge can be applied to heal the energy fields around us today.

Chapter Fifteen

Healing the Energy Field

In the previous chapters I have described the damage to the energy fields surrounding us. This damage results from the enormous increase in electromagnetic radiation with its associated torsion fields. These fields, we are told, are capable of making permanent changes in all biological organisms. Damage has also been caused by environmental pollution and by building in the wrong places. All these factors may account for the significant increase in chronic and degenerative illnesses in recent years. Newgrange was possibly constructed as a healing centre, not just for the people but also for the area because it focussed the earth energy running through it. In this and following chapters I want to explore these subtle energy fields further to see if an understanding of them can help us to repair the damage to both humans and the planet.

I mentioned David R. Cowan (in Chapter 14), and how, as a dowser and geopathic stress consultant, he claims to have healed the energies in many houses and improved the health of the occupants. One of his remedies appears to be Neolithic in origin. He explains how he investigated cup-marks found in stones in Scotland and used this technique to counter harmful spiral earth energies. Cup-marks are shallow depressions believed to have been made by Neolithic peoples and are found carved into stones in Scotland, the North of England and in Ireland. They are about the shape and size of a shallow cup and can be found on large and small rocks, usually in the open, and sometimes underground at sacred sites.

Using dowsing, Cowan claims to have traced the energy associated with a cup-mark and found that it was used to protect a distant ancient dwelling site. I found it hard to believe that a cup mark would be associated with a distant site. Although much of what he says is unsupported by any kind of proof, nevertheless I found it interesting and worthy of further investigation. He explains how you can make a cup-marked stone yourself to protect your house. According to his theory, when you carve a cup-mark into a stone you create an energy field seventy times greater than the stone and this can be magnified still further if the stone is pointed towards a house. This causes harmful energy to go around the house rather than through it and changes unhealthy spirals of energy into healthy ones. He claims a number of successes using this method. In some cases he says that the effect was so strong that the stone had to be temporarily turned off by pointing it away from the house!

As part of my investigations I drove to the Dublin mountains and found a flat piece of granite about 45cm. long and 25cm. across (18x10in.) which was as near as I could get to Cowan's ideal size. Following his instructions,[1] I then set about carving a cup-mark 5cm. wide and 2.5cm. deep. I then carved a circle around it and made a channel facing the sun. My cup-marked stone can be seen in Figure 1 below. This, I understood, would divert any harmful energy around the house, as I was particularly concerned about a nearby mobile phone mast. I then placed the stone close to the house with the channel pointing directly at the house. That night I could not sleep and had to turn the stone to point away from the house.

Figure 1. Cup-marked stone designed to protect the author's house from harmful energies. *(photo K. Comerford)*

I then started experimenting to find the right position for the stone so as to get maximum benefit but not to be so energised that I could not sleep. It is difficult to do divining on your own house as you tend to be too involved in the outcome so I got a friend who divined the best position for it and it is now installed in a flower bed.

It appears that the act of carving in stone, as well as the pattern carved, have a strong energetic effect, as yet not understood. It could be that the patterns carved in the stones at Newgrange were done to set up subtle energy fields for protection or healing or maybe for some higher purpose. Michael Poynder in his book *Pi in the Sky* [2] points out that there are a number of cup-marks in the kerbstone K52 at Newgrange. This kerbstone is at the back of the mound immediately opposite the entrance and is considered to be a point where a number of leylines come together.

When I first read Michael Poynder's book I was a bit sceptical. He describes Newgrange as being "the central powerhouse" of an earth energy grid that runs from the Great Pyramid up through Stonehenge and on through Newgrange to Carrowkeel in Co. Sligo. This is also mentioned by the archaeologist Geraldine Stout in her book which I referred to in Chapter 9. In her comprehensive review of all aspects of the area she includes the so-called "New Age" beliefs and includes a diagram from Poynder's book showing a leyline running straight from the Great Pyramid through Rome and Paris to Knocknarea in County Sligo.[3] However, I have difficulty with this. When the diagram is drawn showing the correct location of the Great Pyramid and using a correct representation of the earth as a globe, it can be seen from Figure 2 below,that if a straight line is drawn from the Great Pyramid through Rome, it goes south of Paris and misses the British Isles completely. I mentioned in Chapter 8 that maps in atlases often use a cylindrical projection which has the effect of enlarging the areas closer to the poles. Thus the alignment claimed is just an accidental outcome of the projection used.

The earth energy grid is stated by Poynder to be composed of hexagonal earth "stars" which stretch across the planet in a connected pattern. These are described by Clive Beadon a respected dowser as "a six-pointed star within a circle, each unit touching the next and linked together by the major energy line running through the centres." He goes on to state that "these patterns, sometimes like the acupuncture lines of the human body, are normally in balance but they can be diverted or destroyed and the resulting disharmony and uneasiness affects people in the area." [4]

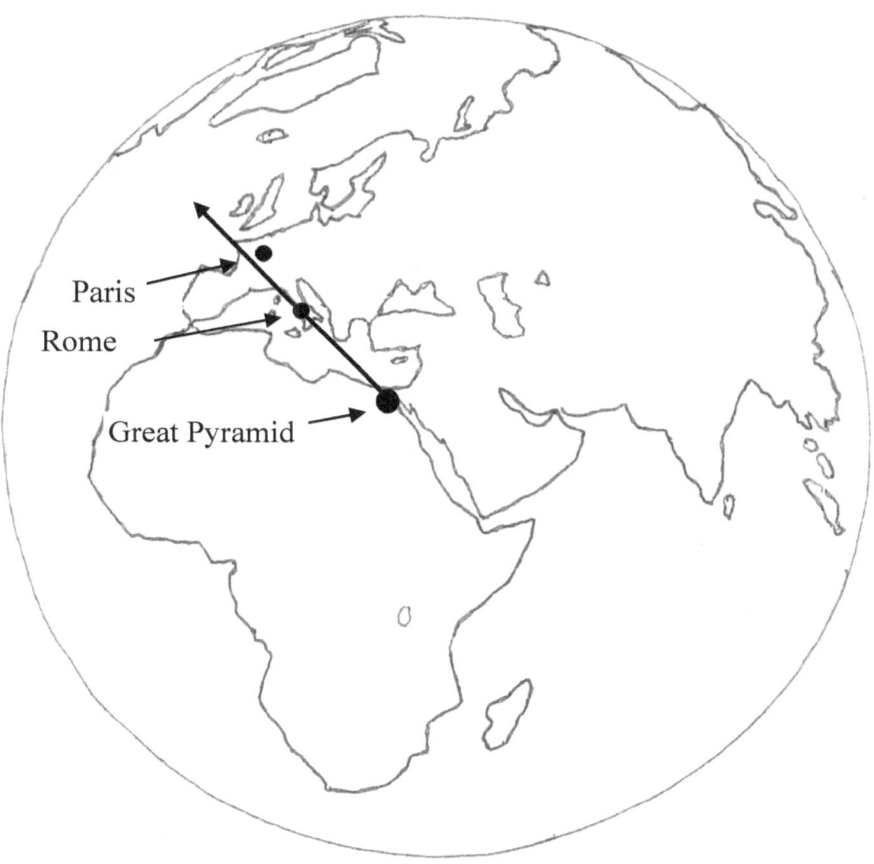

Figure 2. The "leyline" from the Great Pyramid through Rome does not pass through Paris and misses Ireland completely. Newgrange cannot be said to be on a straight leyline from the Great Pyramid and passing through Rome, Paris and Stonehenge.

Wing Commander Clive Beadon D.F.C., a war hero who died in 1996 aged 77, was renowned as a dowser and became vice-president of the British Association of Dowsers. In correspondence with a dowser called Michael Guest, Beadon writes "I think we can say with some certainty that in the distant past mankind knew how to manipulate the various types of earth force lines. He knew how to focus them on to chosen points using specially selected stones which he placed upright (Standing Stones), and some specially grouped which we call Dolmens. With these aids he was able to bring a certain type of power to his stone and wood circles." [5]

Beadon was also well known for developing the "Spiral of Tranquillity", a device designed to correct the earth's unbalanced energy lines within its immediate vicinity. It was intended primarily for use in the home and claimed to be able to restore balance to an area of about 50 square metres. As well as dealing with geopathic stress it was also claimed to be beneficial in dealing with harmful electric fields. I ordered one from Wing Commander Beadon in 1995 and have used it beside my photocopier and computer in an attempt to reduce the amount of electromagnetic pollution, particularly from the positive ions emitted by the photocopier.

Figure 3. Wing Commander Clive Beadon's Spiral of Tranquillity
(photo K Comerford)

The device contains two spirals made of copper wire cast into an acrylic block with a series of coloured gemstones. Clive Beadon's work on these devices has been continued by Michael Poynder who has developed a smaller version for attaching to mobile phones. This has been tested by Dr. Roger Coghill of Coghill Research Laboratories whom I mentioned in the previous chapter. The test report showed that the device had a significant protective effect on human blood lymphocytes when exposed to mobile phone radiation.[6]

Wing Commander Beadon's Spiral of Tranquillity is claimed to create a vortex which restores the original form to a broken earth star. Since most of us live in places where development has taken place over the years without understanding the possible disturbances to the earth energy, it is claimed that many of the original earth stars that existed in the time of Newgrange have been damaged. Regrettably the subject of earth stars seems to be unsupported by any kind of physical evidence and one is relying on the word of dowsers concerning what they have found. I have seen dowsers express differing view on the positions of leylines and earth stars and while this gives rise to a general scepticism, dowsing does seem to work but on a more local level and I will return to this subject later in this chapter.

When considering the possible effects of harmful energies there are many practical steps we can take using accepted scientific principles. We could press for higher standards than those at present in force for exposure to electromagnetic fields (EMFs). These standards which have traditionally been based on the possible heating effects of human tissues by such energies are quite arbitrary and vary widely from country to country. The World Health Organisation is pressing for an agreed framework for the establishment of standards for EMFs.[7] Harmful effects may be found at much more subtle levels of our functioning, both physical and mental, and Dr. Becker's research into the effects of weak electric and magnetic fields on living organisms confirms this. He claims that his work was strenuously opposed by the electricity and telecommunications industry backed by forces within the US Department of Defense.

If we cannot in the short term reduce the radiation emitting from the electrical equipment around us, we can at least reduce our exposure to it by electromagnetic screening. Since we spend one third of our lives in bed, it is essential that we protect our beds from unwanted electric fields if we are in an area where these fields are strong such as areas close to high voltage transmission lines, electricity substations or radio transmission towers. Some protection can be gained by using an electrically conducting curtain like a mosquito net or an electrically conducting under-blanket. This kind of blanket is specially made for this purpose and has nothing to do with electric blankets for heating.

One should never sleep with an electric blanket switched on as this creates a low level electric field over a long period. Even if the field is weak, long term exposure can do serious damage. A study of the use of electric blankets by pregnant women found a much higher incidence of miscarriages than in non-users. In another study of 1,583 female users of computer visual display units (VDUs) it was found that those who used computers for more than 20 hours per week had twice the rate of miscarriages compared with other female workers who did not use computers.[8] Fortunately, most VDUs are now being replaced by LCD (liquid crystal display) screens which are safer.

Other methods of screening that can be used in the home include coating walls with aluminium foil or using nickel screening paint. Some walls made from gypsum panels are already coated on the inside with aluminium foil as a thermal reflector and vapour barrier, but aluminium is not considered a suitable material according to DeMeo. Effective screening requires some technical knowledge and the use of measuring instruments.[9]

As part of my researches into methods of healing earth energies, I attended a weekend workshop run by the Irish Society of Diviners. The venue was a beautiful old house in County Meath close to many ancient sites. The course was given by Alanna Moore an Australian geomancer who is founder of the New South Wales Dowsing Society and has twenty years experience as a dowser working to correct harmful energies in the environment.[10] Geomancy is the ancient art of divination by throwing a handful of earth in the air and reading how it falls on the ground. In modern times the term is more widely used to cover dowsing and the healing of earth energies.

Alanna started by giving us a short course on dowsing and explained the different forms of energy we might find. These are earth energies, sky energies and artificial energies. Earth energies include spirals or vortices, underground streams, geological faults and places where burials or massacres took place. This interested me as I had previously been told by a psychic that there had been a massacre in the area of my house and that there were a number of spirits hanging around. I found out later that my house in Rathmines could be on the site of the Battle of Rathmines which was fought on August 2nd, 1649 and in which the Parliamentarians of Dublin defeated the Royalists thus permitting Oliver Cromwell to begin his massacres in Ireland. I related (in Chapter 13), the comments of Rupert Sheldrake who states that morphic resonances of a disturbing quality can be experienced in places where murders have been committed.

I was interested to learn from Alanna that leylines are sky energies running from hilltop to hilltop. At least this was her interpretation. What I previously understood to be leylines in the ground, she called "dragon lines". The artificial energies are various man-made forms of pollution such as those caused by power lines and radio transmitters, but they can also include energies inside a house due to pollutants and 'sick building syndrome'.

We experimented with dowsing and most of us found a negative energy running across the front of the house. Alanna said that in her opinion an energy such as this could cause people to feel unwell or confused in this area of the house. Then some of the people at the course said that when they entered the house for the first time the night before, they could not remember what to say and were lost for words. Others then recalled that one course participant sitting in this area the night before, had left the course feeling unwell. This may be a bit subjective but what followed was more convincing. We were told that the negative energy was caused by an underground stream which entered the house from the East. A power tower was made from basalt rock dust in a plastic pipe and placed outside the house above where the stream was said to enter. Once the power tower had been put in place I could no longer detect the energy and most of the sixteen people present had the same experience.

Power towers are well known in Australia and hundreds of them have been installed on farms. Many of the farmers claim that they have experienced improved crop yields and a reduction in extremes of climate. The towers are said to function in the same manner as standing stones in that they act as energy concentrators drawing in good energy and overcoming the effects of bad. Alanna decided to erect a larger power tower in front of the building to improve the overall energy of the area. It was made from a piece of plastic pipe about 15cm. (6 in.) in diameter and over three metres (10 ft.) long. A hole was dug in the ground about 60cm. (2 ft.) deep and the pipe was placed in this and the soil replaced around it. The pipe was then filled from the top with basalt rock dust which had been obtained from a quarry in County Wicklow. A ceramic pot was placed on top. This power tower is illustrated in Figure 4 below.

Figure 4.
Power tower in
County Meath erected
and decorated by
course participants.

(photo K Comerford)

The beginnings of a scientific explanation started to emerge when I learned that basalt rock is paramagnetic. Paramagnetism is a weak form of magnetism where rocks become magnetic on contact with a magnetic field but do not hold the magnetism when the field is removed. Having learned this fact I found that one of the experts on this subject is Professor Philip Callahan. I purchased his book on paramagnetism as the next step in my researches which are described in Chapter 16.

There is now a considerable body of knowledge in detecting and neutralising harmful energies. However much of this knowledge is not being put to use because the effects of these energies are not fully recognised. We need to remove the bias that prevents any serious research being carried out on these energy fields. We are not taking advantage of the possible benefits of orgone accumulators and the other energy devices which I have described in this chapter. However, the energy practitioners and others promoting subtle energy and healing often leave themselves open to criticism. In researching this subject I have frequently come across claims made on very little evidence. Regrettably many people working in this field have not had the benefit of a scientific training and often confuse opinion with fact. If the bias against this subject was removed, better quality research could be carried out and we could have more confidence in what is published.

It is possible that by understanding subtle energy fields we could come up with remedies for many of our present personal and planetary health problems. Let us look at one example where we know that Nature can be easily harnessed in our favour. That is the planting of trees which is known to improve the quality of the environment in cities.

Trees take in carbon dioxide and give out oxygen thus purifying the city air and reducing global warming. They collect pollutants from traffic exhaust in their leaves, which they then shed, enabling the pollution to be collected and removed. They give shade in summer helping people to remain cool and comfortable. In winter they provide shelter absorbing wind energy and reducing storm damage. They absorb excess water, preventing flooding, and they encourage the build-up of soil and humus, allowing other plants to grow and provide a habitat for small animals and insects. Planting more trees and grass in cities could improve orgone accumulation and might even reduce the problems of deprivation in so-called concrete jungles. This is because trees grow according to unseen energy fields in Nature, and the presence of trees strengthens these fields which could help to counter the unnatural fields produced by electromagnetic interference, by geopathic stress or by stress in society. A simple policy of requiring all vacant sites to be cleared and planted with trees and grass could produce huge benefits.

A study in Manchester suggests that increasing the tree cover by 10% could reduce the temperature by 3-4 degrees. Yet city authorities are cutting down trees for reasons of health and safety. Examples cited are the danger of chestnuts falling on children or obscuring CCTV cameras. Very few high trees are now planted. Germany has 25% high tree cover, France has 26% and Italy 24% compared with Britain which has 4.5%.[11]

In this chapter we have seen that there is evidence that stones can improve local energy fields and that the effect appears to be magnetic. The energy field of Newgrange was apparently enhanced by the later addition of a circle of standing stones. A modern equivalent in the form of power towers filled with paramagnetic rock has been found by farmers in Australia to be effective in improving crop yields and reducing extremes of climate. In the next chapter the subject of magnetic energy is addressed.

Chapter Sixteen

Magnetic Attraction

Magnetism is a subject with which most people are familiar but which very few people really understand. You might like to test your knowledge by trying to answer the following questions:

What is the basis of magnetic attraction? How is it that some things are magnetic and others are not? If something is attracted towards a magnet where did the energy come from that caused it to move?

In view of what I have been saying earlier about spirals and spin you will not be surprised to learn that magnetism is produced by the spin of the electrons in a substance. There are electrons spinning around the nucleus of all atoms and these are electrically charged. When an electric charge is moving it is called a current and when an electric current flows it produces a magnetic field. On a large scale there are powerful electro-magnets which are constructed by winding coils of copper wire around a core of soft iron. They are used for picking up scrap metal or for sorting iron out from other non-magnetic materials. They also form the basis of electricity transformers, electric motors and generators. On a small scale the spinning electrons in atoms cause minute magnetic fields. Due to the fact that electrons tend to pair up with other electrons of opposite spin, they cancel each other out and so most substances do not exhibit strong magnetic effects.

However some substances, notably iron, have unpaired electrons and these spontaneously align to produce a magnetic force in one direction. This is a quantum mechanical effect and is called ferromagnetism. Other substances which do not exhibit a strong magnetic effect can nevertheless be magnetised by a magnet although they do not become permanently magnetic as iron does. This phenomenon of weak temporary magnetism is called paramagnetism. The regions of a substance which are magnetised are called magnetic domains. These magnetic domains are forced to line up under the influence of the externally applied magnetic field. Another category of substances exhibit a weak repulsive force in the influence of a magnetic field and are said to be diamagnetic. Examples of paramagnetic substances are aluminium and many minerals including most stones.

Diamagnetic materials include water and most organic materials. Substances can be magnetised by bringing them close to strong magnets but they can also be magnetised by hammering them as this also has the effect of lining up the magnetic domains.

Magnets have been known since the earliest times and have occurred naturally in the form of lodestones composed of an oxide of iron called magnetite. However, not all magnetite is found in a magnetised state so how these lodestones became magnetised is not clear. One theory is that they were struck by lightning and that the intense electric current turned them into permanent magnets. This raises the question whether magnets contain energy as presumably the magnetising of them by hammering, by electric currents or contact with other magnets appears to impart some magnetic energy to them. However, this does not provide a complete explanation as magnetic energy appears to have some unusual properties which are not fully understood. For example, a magnet will repeatedly attract a piece of soft iron without any apparent loss of strength as evidenced by magnetic catches on cupboards which seem to last indefinitely. The question of where magnetic energy comes from is still a subject for debate among scientists. The fact that it is recognised as a quantum-mechanical process gives rise to the possibility of energy transfer from the vacuum.

Edward Leedskalnin who constructed Coral Castle had his own version of physics and published his theories of magnetism.[1] Perhaps a conventional science education is not the ideal starting point for understanding his theories as I found them difficult to follow. One example is his Perpetual Motion Holder. He claimed that if you took a horseshoe magnet and put a soft iron keeper across the ends, you could store energy in this magnetic circuit indefinitely. An experiment is described where an electric current pulse is fed into a coil wound around the magnet. This causes a magnetic current or flux to flow. The power source is then removed. After a long time if the keeper is removed and the magnetic circuit is broken, a bulb is lit by the stored energy. This is demonstrated at www.leedskalnin.com. I wondered if energy was required to break the magnetic circuit? If so then what is demonstrated may be simply the energy produced by the change in magnetic flux when the keeper is removed.

I mentioned in Chapter 5 that Professor Myron Evans has claimed that energy can be extracted from the vacuum by the phenomenon of resonance, the tendency for oscillating or vibrating elements to synchronise and transfer energy between them.[2] An oscillating or vibrating circuit is a key aspect of US patent 6,232,718 which was granted to Thomas Bearden and his colleagues in 2002 for a "Motionless Electromagnetic Generator" This patent describes a small prototype generator and claims that it requires an input power of 14 watts to drive its oscillating circuit but then gives an output of 48 watts. In other words its power output is 3.4 times greater than the power input. Bearden claims on his website that the latest version of the device now produces 100 times more energy output than input.[3] It is said to work according to the puzzling Aharonov-Bohm effect which Evans claims to have explained.[4]

An Irish company called Steorn has claimed that it has discovered how to make free energy from magnets. The engineers who run this company say that they were surprised to discover this effect and I was fortunate to obtain an interview with Sean McCarthy the managing director of Steorn in June 2008. He told me that as engineers their aim was to develop a new technology and not to prove new principles in physics. He explained that Noether's Theorem states that the principle of conservation of energy is only true for systems that are time invariant.[5] From their experiments he believes that magnetic energy is time variant i.e., it can be delivered at different rates.

As an example he showed me a test rig where a magnet was mounted on the circumference of a wheel and when the wheel was rotated, the magnet came under the influence of a stationary magnet. By plotting a curve of the torque versus angle I was shown on a computer screen that the energy imparted to the wheel as it approached the stationary magnet was equal to the energy subtracted when it had passed the stationary magnet. However, when another stationary magnet was interposed so as to disturb the magnetic field in a position where it only came into play when the rotating magnet was moving away from the fixed magnet, it could be seen that the energy imparted was greater than the energy subtracted.

Steorn has advertised for scientists to test their ideas and 5,000 scientists applied. Of these they have contracted with twenty one scientists who have been evaluating their technology over two years and, according to McCarthy, none of them has yet been able to prove them wrong. He said that one well known scientist publicly accused them of fraud even though he knew nothing about their technology which had not been published at the time of writing. McCarthy believes that his company has the luxury of not being scientists and not having to prove anything to the scientific community. "We have no credibility", he says, "so we have nothing to lose". He has managed to convince a sufficient number of investors and is making the technology available for license to other engineering companies.

Figure 1. Steorn Orbo demonstration in December 2009. It can be seen that the unit is mounted on a projecting sheet of clear acrylic, showing that there are no external connections. *(Photo K. Comerford)*

During December 2009 and January 2010 the Steorn "Orbo" technology was demonstrated to the public in the Waterways Centre in Dublin. The photograph above shows one of three units mounted on a sheet of clear acrylic to show that there were no wires feeding it. In the demonstrations

Sean McCarthy claimed that the device did not produce a back electromotive force, a type of electrical reaction force, the electrical equivalent of Newton's laws of motion which require a reaction and echoing Laithwaite's claim (Chapter 4). He also claimed overunity, i.e. that the device produced more power output than the power fed to the input from a battery.

The two free energy devices described above appear to make use of the "two elephants principle". The vacuum has been described as two elephants pushing against each other.[6] The trick is to find out how to remove one of the elephants. In each case it appears that a small amount of energy is used to divert the power of a magnet thereby releasing free energy that was otherwise being held in check. It is possible that what we are witnessing here is an unbalancing of a magnetic equilibrium which allows the release of the otherwise hidden torsion energy.

Now that we have learnt something of the puzzling possibilities of magnetism let's see if we can apply our knowledge to learn more about ancient sites. When I read Dr. Phil Callahan's book on paramagnetism,[7] I found out that he had spent a considerable time in Ireland during the Second World War as a radio technician working on low frequency radio ranging equipment. He had gone on to become an entomologist and this strange combination, a knowledge of radio and insects, led him to understand how insect antennae receive minute electrical signals and to apply the same principle to explain the function of Irish round towers.

Callahan found that insects' antennae were able to tune or resonate with Extremely Low Frequency (ELF) waves as were round towers. An antenna can be made from an insulating or dielectric material. Insulating materials such as stone worked best but semiconductors such as insect antennae and trees also worked. These had the effect of tuning and amplifying various frequencies of the waves. He found that paramagnetic materials such as basalt and granite gave quite strong results and suggested from his observations of ancient Feng Shui gardens in Japan that a combination of paramagnetic and diamagnetic materials enhances the effect.

ELF electromagnetic waves are produced in the earth's atmosphere by a resonance phenomenon. Radio waves even of low frequency can be tuned by a resonant cavity which is related to their wavelength. The space between the earth's surface and the ionosphere, a reflective layer high in the atmosphere, acts as a resonator to ELF waves which have a fundamental frequency of 7.83 Hz with additional resonant harmonics at 14, 20, 26, and 33 Hz. Strangely, these frequencies exactly match the EEG brain frequencies with 7.83 Hz being the frequency of the alpha wave found in meditation. This frequency is called the Schumann Resonance after the physicist Winfried Otto Schumann who first predicted it mathematically in 1952. The resonance is maintained by lightning discharges which are occurring at different places around the world and so the Schumann Resonance propagates as a stationary wave which circles the earth.

Figure 2.
Round Tower at Glendalough Co. Wicklow, an energy enhancing device?.

(Wilipedia Commons/Superbass)

Irish round towers are believed to have been constructed in the 5th and 6th centuries by monks as they are usually associated with the ruins of monasteries. They are considered to have been used as bell towers or as places of refuge where the monks would be safe from raiders. However this theory has been discounted in recent years. The entrances are usually 6 to 12 feet above the ground and the theory was that the monks would bring their gold chalices and other valuables with them into the tower and then withdraw the ladder. However the towers contained wooden floors and there was not enough vertical room to stow a ladder. It has also been pointed out that the towers would act as excellent chimneys and that the monks would be incinerated if the raiders succeeded in putting burning materials in through the doors.

Callahan believes that they were used to improve the energy of their location and that this helped agriculture and healing generally. They were the first power towers and were built because they were highly paramagnetic, usually being constructed out of paramagnetic rocks such as granite. Callahan carried out numerous magnetic measurements at round towers using a special meter he developed, and conducted laboratory experiments where he showed that plants grew better, if planted close to a model round tower which he made paramagnetic by coating it with a sheet of carborundum paper. He plotted the magnetic fields and demonstrated how they were increased in the vicinity of the towers. He also took measurements which showed magnetic effects at the Neolithic site of Loughcrew.

But Callahan's main interest in his book on paramagnetism was in the promotion of better techniques for growing plants. He showed that plants grow best on paramagnetic soils and claimed that insects do not attack healthy plants grown on healthy soil. He advocated the spreading of basalt and other paramagnetic rocks in powdered form as a substitute for chemical fertilisers citing examples where soils have recovered quickly after the eruption of volcanoes, the spread of volcanic ash revitalising the soils and improving crop yields.

Having studied Callahan's book I began to see how its principles could be applied to the energy of Newgrange. Firstly I thought about the standing stones and their possibilities to increase the energy field in the mound. Then I thought about the kerbstones and their designs which had been chiselled, the hammering possibly increasing their magnetism. Next I thought about the alternate layers of paramagnetic stones and diamagnetic sods within the mound and also about the quartz at the entrance.

Newgrange was originally surrounded by a circle of possibly 35 to 38 standing stones of which only twelve survive today. The diameter of the stone circle is 103m. as compared with the mound diameter of 85m. and it is suggested that the stone circle was a later addition. The stones are made of greywacke a kind of sandstone which Callahan lists as having a medium value of paramagnetism. Thus they would have enhanced the energy of the mound and the area around it. It is interesting to note that other materials used in the mound such as granite, which have higher paramagnetic values were imported possibly from Wicklow thirty miles to the south or from the Cooley peninsula thirty miles to the north.

In the previous chapter I described the work of David R. Cowan on cup-marked stones and his statement that by making a cup-mark in a stone, you increase its energy by seventy times. He does not explain this but it could mean that the hammering, which is known to align magnetic domains, increases the paramagnetism of the stone. I certainly confirmed the power of such stones as the one I made was so powerful it affected my sleep!

The kerbstones at Newgrange and more notably at Knowth are covered with designs ranging from spirals to waves and other features one of which is claimed to be an ancient calendar. It is possible that the chiselling of these stones increased their magnetism but I am not suggesting that this was the only purpose of the designs. It was one purpose and the designs themselves could also have had symbolic meanings, particularly the spirals and waves.

I was fascinated to find that the alternate layers of organic and inorganic materials were diamagnetic and paramagnetic. Also the Japanese idea of combining the two suggests some form of amplification of the energy. This supports the theory which I put forward in Chapter 7 that the mound acted as a kind of capacitor storing and increasing the level of the energy. Capacitors connected together act something like batteries in series. If you want a higher voltage you connect a number of cells in series with each other and the voltages of the individual cells are summed to give the total battery voltage. However, I am not claiming that this is a form of paramagnetic energy, rather that the paramagnetic field enhanced the overall electromagnetic field and the associated torsion energy.

I had always wondered about the extensive finds of imported quartz rocks at the entrance to Newgrange. These are not found anywhere else in any quantities and it is clear from the excavations that they were never part of the internal mound materials. However Callahan shows that such structures have a weaker energy field at the eastern side. The placing of substantial amounts of diamagnetic quartz at the entrance facing the rising sun in mid-winter would have had an amplifying effect when combined with the paramagnetic stones inside the mound itself, thus enhancing the energy at a time when the solar energy was at its weakest. On my visit to Knowth I noted that amounts of quartz were present on the ground at both the east and west passage entrances. The amounts were much smaller than at Newgrange and this adds to my belief that Newgrange was a later and more advanced construction.

Callahan also talks about getting minimum energy signals at ground level and speculates that this was why the doors of the round towers were placed much higher up. He states that in constructing radio ranging stations in Japan after the war they often had to construct a grid about six feet off the ground before the radio station would work. This is because there is a null point where the radio wave is reflected off the ground. I like to think that it also explains why the kerbstones were used. I mentioned before that they may have been insulators but they could also have served to raise the top of the mound into a higher signal strength area.

Trees are claimed by Callahan to be antennae which attract energy although they are made of organic material and therefore by his own definition diamagnetic. He seems to be saying that the energy is attracted to them as antennae. However the round towers are also antennae and are paramagnetic. I would like to be able to say that I agree with everything Callahan says but some of his comments and results are confusing. He says also that concrete is paramagnetic so you would expect it to attract good energy but we know that concrete jungles are places with bad energy. Perhaps concrete is beneficial when it is used vertically as in a building but harmful if it covers the ground thus preventing the beneficial effects of soil, grass or trees being received. This is borne out by a consideration of solar energy in the form of electromagnetic radiation as it falls on the earth.

Figure 3. Standing stones, round towers and power towers concentrate electromagnetic energy from the sun as do orgone accumulators such as Newgrange.

In Figure 3 solar energy in the form of electromagnetic waves is falling uniformly on an area of the earth. Interposing a standing stone or power or round tower concentrates the energy in the area of the tower because of its paramagnetic properties. The magnetic component of the energy flowing downwards through the stone or tower produces a torsion component called a magnetic vector potential. Thus the energy is concentrated and a torsion component is produced which may enhance the healing energy in this spot or overcome otherwise harmful energies. The paramagnetism of the stone or tower may also help to tune the incoming energy to resonate in sympathy with the Schumann Resonance.

This principle can also be applied to an orgone accumulator. Since metals are more permeable to magnetic fields than stone it follows that Reich's orgone accumulator composed of alternating sheets of metal and wood would be more effective than a Neolithic mound of equal size. Metals having higher paramagnetic values would be able to concentrate energy better than stone. Orgone accumulators of relatively small size appear to be quite powerful and the user is cautioned against overexposure to orgone energy. It should also be noted that some geopathic stress consultants use metal rods instead of power towers. This is called "Earth Acupuncture". What is important here is to learn the principles of subtle energies and to be able to apply them correctly.[8] This means that it may be possible to use metals which could be more effective than stone and could be used in smaller quantities. While stone may be effective, it was the best material that was available in the Stone Age and it might not be necessary to construct something as massive as Newgrange to get the same result today. In any event, great caution should be exercised as the presence of metals such as aluminium can also have a negative rather than a positive effect.[9]

It is clear that we are only beginning to scratch the surface of this subject and it will be necessary to carry out a considerable amount of research before we can understand the principles behind this newly rediscovered energy. Until we know how it works we must be careful about applying its principles as we may do more harm than good. We could be like the doctors of the nineteenth century who did not know that it was necessary to wash their hands between examining patients. The next chapter will introduce some of the known and safer methods of using our present knowledge.

Chapter Seventeen

Applications of Subtle Energies

In this book I have described my researches into subtle energies which have been triggered by my interest in Newgrange, alternative science and spirituality. The book is centred on Newgrange but has described a number of new developments and ideas which are essentially holistic. In other words, in looking into Newgrange and its mysteries we find answers to questions that have much wider applications because they delve into the laws of nature at a very fundamental level. The resulting ideas can have applications at the level of the individual, of society and of the environment. In this chapter I am going to concentrate on the level of the individual.

Before going on to discuss the applications I want to state that in introducing various ideas I am not claiming that they are all essential components needed to understand the universal energy field (UEF). Many of the things I have described are interesting but may or may not turn out to be essential to the working or application of this knowledge. I am sure that many scientists will be sceptical of some of the things I have described and pick on some aspects as speculative and unproven so let me start by saying that you do not have to accept all the points that I have made. However, I believe that there are enough supporting theories to cause any scientist who is open and objective to look again at some of the ideas which had previously been dismissed as unscientific.

I would like to issue a word of caution about how science is to be applied to the ideas in this book. On the one hand the scientific method requires that experiments must be repeatable by any qualified person, while on the other hand, quantum physicists tell us that the observer is part of the experiment. How can an experiment be repeatable if the observer is a different person? The intentions and expectations of the researcher concerning the outcome are not being taken into consideration, although we now know that the power of human intention or emotion can affect the physical state of matter. We see this in the experiments on the curling up of DNA, from the REG trials in Princeton and from Emoto's experiments with water.

The scientific revolution has created the popular view that "seeing is believing". Scientists have often expressed the view that if they cannot measure something it does not exist. But when we study scientific papers we find that scientists are actually not saying this at all. They are saying "we carried out measurements using certain equipment and the readings did not prove that the suggested effect was present". This is not the same as saying it does not exist. All that they are saying is that they did not find it. There is a well-known saying in research circles that "absence of evidence is not evidence of absence".

Regrettably there have been cases of scientific fraud reported which go to prove that some scientists are not just seeking to find the truth but seeking to confirm a particular result which will benefit them in some way. Even the most honest of scientists will agree that it is hard to avoid a preconceived view of the outcome of an experiment. Unfortunately we must accept the fact that this may actually affect the result. My view is that in cases of subtle energies, the effects are quantum mechanical and highly sensitive to mental intention. The test of repeatability must also be accompanied by a proven lack of bias. How this can be achieved is something that needs to be considered.

I have stated that the subtle energy I described can be applied at the level of the individual, of society and of the environment. At the level of the individual we find evidence of healing coming from a number of sources. Let us look first at the aspect of stress reduction and the use of meditation techniques. Here we find that there are so many different techniques that it is difficult to conduct tests and draw conclusions. However, Transcendental Meditation was taught to millions of people in the 1970s and 80s and because the technique was standardised it was possible to carry out tests with statistically valid samples. Over 500 studies carried out at more than 200 universities in 30 countries showed substantial improvements in stress levels and in mental functioning in the individual and in society.[1]

Meditation appears to connect the individual to a more universal and holistic energy field. Many scientists now believe that the brain is a quantum-mechanical device which works on a holographic principle, i.e. a principle that makes use of the property of interference between waves.[2]

Pictures taken using this principle are called holograms and one of the interesting properties of a hologram is that if you cut out a piece of the picture and shine a laser on it, the full picture emerges. This demonstrates the holographic principle that holograms are holistic or 'whole', i.e. each part contains the whole picture. If the brain works holographically then it must have access to information from the 'whole', from the UEF.

I mentioned (in Chapter 12) the fact that some scientists now believe that consciousness is not confined to the brain at all,[3] and it has been suggested that memory is not just a question of accessing something in the memory banks of the brain, but of using the brain as a receiver to connect to a universal memory bank.[4] This allows one to conclude that we can access this UEF by using our thoughts. If this is so why don't we do it consciously instead of unconsciously? The reason is simple. We live in a society where we have lost the ability to use these powers. We can express thoughts, intentions or desires, but they are fighting against the prevalence of background noise caused by the stressful world we live in. Even when we relax and try to rest we do not achieve anything like the level of quietness of brain function that has been measured in meditators. The physiology reacts equally to our thoughts as to real situations, and we are unable to let go of our mental conditioning which carries over stress from past events into the present moment. So our mental energy field is also damaged and this may have started with the loss of the partnership society so long ago.

I believe that the people of Newgrange were able to live in the present, unconcerned about the past or the future. They lived in a society where worldly goods were shared and where one did not have to worry about being attacked or robbed. It is known that the climate at that time was warmer and so food was plentiful and shelter from the cold was not an issue. In such a society there would have been a very low level of stress by our standards. In fact we could say that some of these people were so relaxed that they were in a state of meditation without having to meditate. Their brainwaves would have been mostly in the 8 Hz range as is found in young children. They were also tuned in to the Schumann Resonance and therefore totally in tune with the planet and its energies.

This was their reality. To recreate it we need to tap into this energy field by taking time out. It is necessary to meditate regularly until one reaches a state described in India as cosmic consciousness, a permanent meditative state while still fully awake. But how can we reach this state while living in such a chaotic world? We have to do it by performing a reset. This is what you do when your computer has developed a mind of its own. You press the reset button and it drops whatever misprogramming it has picked up and restarts afresh.

Our misprogramming can be seen as defects in our energy fields. For each of us our part of the UEF, our Human Energy Field (HEF) is damaged by past incorrect thinking or programming, which has caused us to continually make wrong choices. Let us take a detailed look at the HEF so that we can learn how to repair the defects and reset our lives. We can examine the HEF from two different angles. The first is the scientific approach which shows us how this field is being measured and understood. This is leading us to develop equipment for treating imbalances using scientific devices, mostly electronic. The second, the clairvoyant approach, is leading to methods of balancing the HEF by tapping into energy from the UEF through the body of the healer.

Science has made vast strides in recent years in detecting very weak energy effects. Living tissues are conductors of electricity, and signals which can be detected quite easily are sent between organs such as the brain and the heart. Electric currents in the body produce magnetic fields which are able to give much better quality information than the electric fields previously measured. Instruments called magnetometers are now used to measure the HEF. The most sensitive of these is the SQUID (Superconducting Quantum Interference Device), a device so sensitive it can measure the minute magnetic fields caused by electric currents flowing in the body. These fields are 100,000 times weaker than the earth's magnetic field.

The HEF has traditionally been called the aura and the earliest scientific method of aura measurement was Kirlian Photography, referred to in Chapter 12. This method placed objects in a high voltage electric field and an image was exposed on a photographic plate under the object. The method was used to show the energy field of various objects such as leaves and even showed the remaining energy field if part of the leaf was removed. It is not suitable for recording the full human aura and some scientists have cast doubt on it because it seems to give aura readings for organisms which had died.[5]

Aura photographs are often on offer at Mind Body Spirit exhibitions and many people think that these are Kirlian photographs. In fact they are obtained by picking up signals from the two hands of the subject and projecting them onto a Polaroid photograph of the head of the person so that they appear as an energy field around the head. They do not represent the aura around the head but are a reading of the aura of the hands and can be read by a skilled practitioner with useful results.[6]

HEART BY-PASS OPERATION

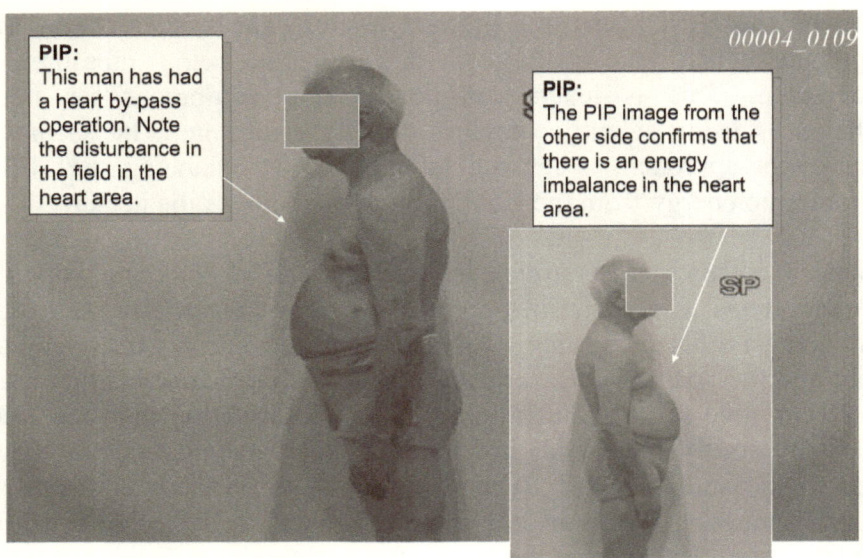

Figure 1. An aura measured by Polycontrast Interference Photography. Note the distorted energy field in the heart region after a heart by-pass operation. It is easier to see in the original colour slide.
(by kind permission of Harry Oldfield)

Another method of aura measurement is Polycontrast Interference Photography or PIP which was developed by Harry Oldfield, a successful healer and inventor in London.[7] This technique detects interference patterns between the light illuminating a subject and the person's energy field. A TV camera feeds a signal into a computer where it is analysed and appears on the computer screen with the colours of the aura. A PIP scan showing an aura appears opposite. The technique is used in conjunction with a method of measuring the aura using sound which is called the Electro-scanning Method or ESM.

ESM uses a sensitive acoustic sound meter to measure bodily vibrations in the audio range. These are associated with the positions of the chakras or subtle body energy centres, previously known in the East on an intuitive basis. Harry Oldfield has shown that these actually exist and can be measured physically. He has also developed a method of healing using acoustic generators which he claims can feed back healing signals at the correct frequency to rebalance the chakras. These acoustic signals are applied to the body by tubes of crystals which use the piezo-electric[8] effect to transmit the healing energy. It is called Electro-Crystal Therapy.

The chakra energy system is a method of describing the HEF which goes back to the Vedic tradition. Chakra means a wheel in Sanskrit and the chakras have been seen clairvoyantly as rotating vortices. When talking about chakras, people mainly refer to the seven principal chakras or energy centres.

The seven chakras are: first, the base chakra which is at the base of the spine and is concerned with survival. It is seen as a red colour. The second or sex chakra is located above the genitals and is concerned with reproduction and is orange. The third is located in the solar plexus or stomach region and is concerned with power or control. This is yellow. The fourth or heart chakra is concerned with love and not surprisingly is located at the heart. Green is considered to be the colour of heart energy.

The fifth chakra, which is located at the throat, is blue and is concerned with creativity and communication. The sixth or third eye is located between the eyebrows and is concerned with inner vision or spiritual communication, and is coloured violet. The seventh or crown chakra is above the top of the head and is concerned with contacting the UEF or, in spiritual terms, with receiving divine light or experiencing oneness with God. Its white colour represents the universal white light and includes the colours of all the other chakras. This is also true in physics where white light is made up of all the other colours of the spectrum.

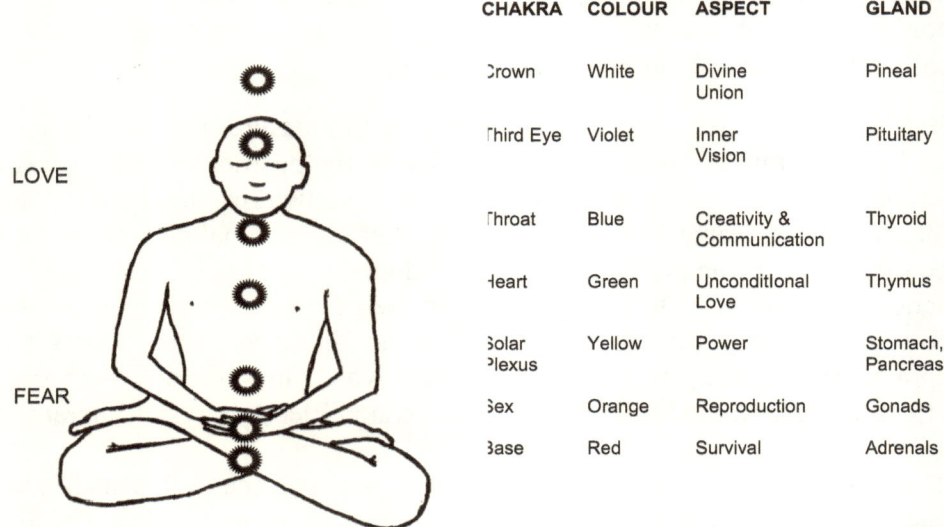

CHAKRA	COLOUR	ASPECT	GLAND
Crown	White	Divine Union	Pineal
Third Eye	Violet	Inner Vision	Pituitary
Throat	Blue	Creativity & Communication	Thyroid
Heart	Green	Unconditional Love	Thymus
Solar Plexus	Yellow	Power	Stomach, Pancreas
Sex	Orange	Reproduction	Gonads
Base	Red	Survival	Adrenals

LOVE

FEAR

Figure 2. The chakras or energy centres in the Human Energy Field (HEF) The lower chakras could be associated with the dominator society and fear and the upper chakras with partnership spiritual development and love..

Although the chakras are detected in the aura, they have direct counterparts in the glandular system of the physical body. This is because the aura or HEF is the etheric or subtle counterpart of the physical body. Illnesses usually show up in the aura long before they manifest in the physical body. The relationships are shown in Figure 2. above. The positions of the chakras and the glands do not correspond exactly.

If we are in danger or experiencing any kind of stress our adrenal glands, which are related to the first chakra, secrete adrenalin and cortisol. Blood supply is reduced in non-essential organs, our heart rate increases and we are primed ready for fight or flight. Unfortunately our modern lifestyle seldom gives us the opportunity either to fight or to flee and we have to restrain our physical readiness, which we usually convert to anger or frustration at our lack of power to control the situation. This affects the third chakra which is related to the stomach and can give us ulcers or bad digestion.

An interesting way of looking at the chakras is to consider that the first three chakras are concerned with fear and the higher ones with love. The first or survival chakra is concerned with fear of death or of not surviving, the second or sex chakra with fear of not producing progeny or of not being loved or supported in old age, and the third with fear of being enslaved or controlled by another person. The fourth or heart chakra is concerned with love and the start of spiritual development and the final three deal with the completion of spiritual development, becoming the creator, developing spiritual communication and experiencing the unity of all creation. It is possible that in moving backwards from a partnership to a dominator society many people have closed down their higher chakras.

The chakras are sometimes described as vortices. People who claim to be able to visualize these subtle forms of energy see the chakras as rotating about horizontal axes in the body. Each one is like a vortex or cone of energy radiating out of the body from front and back. Barbara Ann Brennan is well known as a spiritual healer throughout the world. She found that she had developed a gift which she calls "High Sense Perception" which enabled her to see people's auras and chakras and to diagnose illness and treat it using hands-on healing, which directs subtle energy from the UEF to repair damage to the HEF.

Her story is interesting. She grew up on a farm in Wisconsin and spent a lot of time on her own. Like Victor Schauberger she learned to watch what was happening in Nature all around her and became very sensitive to energy fields. She relates how she could walk through the woods blindfolded and could feel the trees long before she touched them with her hands. She later studied science and became an atmospheric physicist working for NASA. She contends that all illnesses result from some aspect of incorrect thinking and claims to see the effects of negative thoughts, difficult relationships or wrongly held beliefs in a person's energy field. These, she believes, are the precursors to disease. Her work and techniques are described in her book *Hands of Light: A Guide to Healing Through the Human Energy Field.* [9] It adopts a scientific approach in explaining subtle energies.

Most spiritual or subtle energy healing is done by balancing the chakras. The view of healers is that, since spirit manifests as body, by curing the energy system of the etheric body it is possible to get at the root cause of physical ailments. The energy template of the body is its morphic field, the pattern to which the body grows. Thus if we influence this field by our thoughts, we may create our own illnesses through negative thinking. Similarly we can cure ourselves by our thoughts but most people need help because of our mental conditioning. There are many types of energy healing such as Reiki and Therapeutic Touch which consist in laying the hands of the healer on the body and directing the UEF which flows through the healer's hands to balance the chakras.

Scientists have studied healing techniques to test their effectiveness and to find a possible explanation for how they might work. Albert Roy Davis and Walter C. Rawls were working on subtle energy fields and healing as early as the 1930s. Their research showed that energy from the hands of healers flowed in spiral vortices and possessed both electric and magnetic polarities. For example the palm of the right hand was shown to have a magnetic south pole, a positive electric polarity and a clockwise vortex. They also gave an explanation for magnetic fields in the body and suggested that the energy did not originate in the healers but was obtained from the earth's magnetic field by a process of tuning-in or resonance.[10]

Experiments have shown that healers generate a healing radiation through their hands which sweeps through a range of frequencies and often occurs in bursts. SQUID measurements show that energy emitted from the hands of healers has a basic frequency of 8-12 Hertz. This frequency is the same as the Schumann Resonance, a basic resonance frequency of the electro-magnetic field of the earth which was described in the previous chapter.[11] It is also the same as that of the alpha waves found in meditation. This certainly leads to the conclusion that the mind is in resonance with the body and the body is in resonance with the earth. It goes some way to explaining the concept of healing techniques such as Reiki which are claimed to use some form of energy from the hands.

I described (in Chapter 5) the demonstration of subtle energy in the hands where a person is lifted off a chair. The participants are asked to place their open hands on the person's head with alternate left and right hands in a vertical pile. What appears to be happening is that two spiral healing waves one with a right hand polarisation and one with a left hand polarisation are being applied simultaneously. These may be torsion waves and appear to have the effect of reducing the force of gravity.

The alternative science literature contains a number of references to the use of "Caduceus" coils or electrical coils which are connected so that they produce electromagnetic fields which cancel each other out. Two coils are connected in opposite polarity so that the current in one flows in the opposite direction to the current in the other. The theory is that if there is no electromagnetic field, what is left is the associated torsion fields which do not cancel each other but instead may add together to produce a scalar field, i.e. one with no direction. This scalar field could be the source of healing energy.[12]

There are many reported cases of healing using hands-on techniques but since they have not been carried out under controlled conditions they are dismissed as anecdotal. This does not mean that they do not work, merely that the effects have not been measured scientifically. They should not then be dismissed by scientists until proper tests are carried out and without bias on the part of the researchers.

I believe that Newgrange was constructed to concentrate subtle energies which were conducive to healing, meditation and divination. These ancient techniques have been rediscovered in recent years as society begins to question the overuse of the scientific paradigm. We do not have to prove everything according to the scientific method and in many cases, especially those involving subtle quantum mechanical effects, it appears that the scientific method needs to be revised. As stated earlier, we cannot always expect sensitive experiments to be repeatable by different observers with different intentions since it has been clearly shown that the intention of the observer can affect the outcome.

The problems in society such as violence and disease all have their roots in incorrect thinking. The illusion of separateness resulting from the loss of our connection with the UEF causes us to see others as a threat and to focus on our fears. Seeing ourselves as separate causes us to spend our energies defending our individual egos and the positions we have built up in society. However, as we shall see in the next chapter, the ancient technology symbolised by the spiral is now showing us the way to reconnect with the UEF through the power of intention. Seeing ourselves as connected to everyone else leads to the reconciliation of differences.

Chapter Eighteen

The Power of Intention

In the last chapter I attempted to show how the knowledge of subtle energy which was triggered by my curiosity about Newgrange could be used to enhance and explain many techniques which are outside the realm of current science. Healing can be applied to the individual but it can also be applied to society and the environment. In this chapter I hope to show how individual use of subtle energy techniques can heal society.

We can never return to the reality that existed at Newgrange in 3,200 BC and I am not suggesting that we revert to the Stone Age. However, I believe that it is possible that we could achieve a modern society free from domination, violence and terrorism. We can do this by healing our energy fields and those of the planet. We can create partnership energies to overcome the dominator ones. We can do this simply by the power of thought. Our thoughts are made immensely powerful in the state of meditation.

Up to the present, most people have seen meditation as something to use to relax, to overcome stress in their lives. But we have now moved on to a point where we know that meditation can do much more. We can use it to perform the reset I mentioned in the last chapter; to go back to our original programming which existed before we collected the bugs in our programmes - the baggage that we all carry around due to past mistakes, disappointments and wrong thinking. Having reset our mental computer we can restart it to create a completely new reality.

The people of Newgrange were in touch with energies and levels of awareness that we cannot normally contact due to the noise in our environment and in our brains. People who are totally at peace sometimes slip into a meditative state and you may have done this while on holidays in a beautiful place, or sitting alone looking out at the sea or a lake. It is natural for the brain to seek its most restful state, but we no longer live in a natural world and so the brain cannot easily contact this state, that is unless we trick it into stopping its constant whirring.

The various meditation techniques are designed to distract the mind from its normal state of overactivity and to lead it into quieter states. Some techniques such as Transcendental Meditation use a specially selected mantra, literally a 'mind-instrument', to take the mind to a more settled state. Other traditions use divine names or universal mantras like 'OM'. The Buddhist traditions use techniques like following the breath or visualisations. The various methods and benefits of meditation are described by Herbert Benson MD in his best-selling book *The Relaxation Response*[1] and I have also included a brief guide to meditation as an appendix.

Meditation is a process of spiralling in from many thoughts to a single thought and then to no thought. The spiral of Newgrange represents the process of going within, of contacting the inner world. As the attention is diverted from the active outer world it spirals inwards, raising its frequency and thus contacting higher dimensions and states of consciousness. Here, consciousness means more than just being conscious. It means awareness of one's spiritual dimension and bringing that awareness into the body. Professor Tiller gives a formula for consciousness based on information theory.[2] He says that consciousness increases with the frequency and signal power of a system and decreases with the noise level in the system.

The process of meditation meets the requirements of raising consciousness since it has been shown to improve mental functioning, i.e. the ability to think effectively and not to become confused or stressed. It increases the power of thought by improving an individual's ability to focus. Scans of the brains of people meditating show a slowing down of brain waves and a reduction of noise. By reducing noise the brain is able to settle to a very silent state, to go from diversity to unity, from individual awareness to cosmic awareness.

John Hagelin, whom I mentioned earlier, relates the state of meditating to contacting the UEF, a state of "all possibilities" and of infinite energy.[3] Rupert Sheldrake has stated that the power of the field is related to the number of organisms that make it up. There is also an intensity factor related to the power of each individual, to the level of consciousness which he calls the "consistency or intensity of a pattern of thought or intention strongly held". I referred to this in Chapter 13 in connection with the discussions between Sheldrake, David Bohm and the philosopher Renée Weber.[4] She asked Sheldrake and Bohm about the power of a rare individual such as Buddha or Christ. Bohm stated that the power of a Buddha-field came from the wholeness deep within the implicate order. Moving towards Bohm's wholeness or 'higher dimensions enfolded in on themselves' mirrors the terminology of other researchers who describe going to a state of increased symmetry and infinite energy, or spiralling into a higher frequency of vibration and higher dimensions.

| Old Reality | Spiralling in | No Thought | New Thought | Spiralling Out | New Reality |

The spiral – moving from the manifest world by means of meditation to get the mind to a settled state from which a new reality can be created.

The spiral points to the inner way and, while the Newgrange society did not leave any written records, the contemporaneous Vedic society did, and we can turn to Dr. R. K. Wallace's book *The Physiology of Consciousness* [5] for an explanation of the power of a thought acting from a settled mind, which is based on recent research into the ancient Vedic texts. In the diagram above, adapted from one in Dr. Wallace's book, it can be seen how an individual can go through a process of creating a new reality. One starts at the left hand side in the old reality and then spirals in through meditation to a state of no thought thereby accessing all possibilities and infinite energy. A transformation can then take place. A new thought is introduced creating a new reality in the consciousness of the individual who then emerges into that new reality.

This is the vision for the transformation of society and the planet and it can be realised, not overnight, but gradually, by a change in attitude and behaviour. If enough people express the desire for a partnership reality it will come about. If these ideas are expressed from powerful quiet minds it will come about much sooner than we think possible. Some ideas seem to suddenly appear from nowhere. In his best-selling book *The Tipping Point*, Malcolm Gladwell [6] shows how the acceptance of a new idea occurs in a manner which is often unrelated to the amount of exposure it gets. There appears to be a threshold point where certain ideas just take off inexplicably. This is a consciousness or field effect and may be similar to the concept of global consciousness described earlier by Dr. Roger Nelson.

The idea of activating a powerful idea from a settled mind is not unknown in modern society. In fact it forms the basis of hypnotherapy. Many people have a view of hypnosis based on what they have seen on TV or in a theatre. This gives the impression that a person who is hypnotised gives over total control of their mind to the hypnotist. However people under hypnosis are not asleep and are still able to resist any suggestion that is damaging to them.[7] They can, in fact make the suggestions themselves as Valerie Austin, a consultant hypnotherapist, explains in her book *Self Hypnosis: A Safe Self-Help Guide*.[8]

In a hypnotherapy session a client is led to a state of deep relaxation by the therapist. Suggestions are then made by previous agreement, which help the client to create a new behaviour and a new reality for themselves. In my researches, I have experienced both meditation and hypnosis and I could not distinguish between the relaxed states induced by these two methods. The procedure used for relaxing me in hypnotherapy was similar to guided meditations I experienced where a group of people are led on a journey to deeper states of relaxation. However, in hypnotherapy, the therapist then intervenes to make suggestions which help the client to alter their programming and improve their situation. The client is trained to do this themselves in the absence of the therapist and the technique is the same in general terms as that described above which comes from the Vedic tradition.

Hypnotherapy is a powerful and successful technique but is still frowned upon by many psychiatrists and psychologists. This is because there is still much disagreement about and a limited understanding of the workings of the human mind. Dr. David Orme-Johnson refers to this in a ground-breaking article in *Modern Science and Vedic Science*,[9] "It is widely recognised," he says, "that 20th century psychology has not made substantial cumulative progress. Consequently it has not been able to solve the extensive psychological problems of our time, ranging from individual stress to international conflicts". He goes on to describe a new model for psychology based on recognising the ground state of awareness experienced in meditation as the foundation for a scientific structure of psychology which links into other sciences such as physics.

He states a number of new principles which provide a new way of looking at older ideas. The evolution of the universe is not just based on physical laws, he claims, but is based on consciousness which forms the basis of all natural laws. Different behaviours and realities are experienced in different states of consciousness. Nature is not neutral; its purpose is the evolution of life through higher states of consciousness. Objective knowledge alone is not sufficient for a full understanding of natural laws; subjective knowledge must also be recognised. The problems of society can be solved by going to a higher level of awareness from which a complete solution can be obtained.

Unfortunately it seems that, while this new knowledge for the transformation of society has recently become available, it is difficult to get the scientific establishment to approve or promote its introduction. The continual rejection over the years of new ideas which threatened the scientific edifice has not served us well. A new kind of decision-making process is needed in society, one which allows people to express their intuitively held desires and opinions and which is not vetoed by scientific logic based only on objective criteria.

The churches could show the way but, as I stated in *The Christian's Dilemma*, they have so far felt threatened rather than empowered by the new spiritual consciousness.[10] However, many members of the clergy are moving to express more holistic teachings and many go much further in private than they will go in public. If enough people encourage their religious leaders they will feel more confident in expressing a more inclusive and a less exclusive view of religion. Why should people of one church not be welcome in another? What does it matter if they have differences of belief? In a society freed from fear, religious differences could become like cultural differences, something to be celebrated, and if a bit of fusion takes place, why not? This could lead to a more ecumenical society and ultimately to a breakdown of the divisions between Christians and the other major traditions.

Fortunately, we have now arrived at a point where mass communications have enabled this new decision-making process to emerge. The Internet with its access to uncensored information has allowed spiritually inclined people to contact each other and share their opinions and desires. A vast body of rising global consciousness is emerging and gaining power through spiritual, healing and alternative science websites. This information is slowly beginning to penetrate into the mass media and many people now feel quite comfortable discussing spiritual matters in a non-religious context.

The Internet has also facilitated the carrying out of experiments where large numbers of people can express the power of their intentions to change reality. These have been shown to confirm the principles of creating a new reality which I have described above. Lynne McTaggart whom I mentioned in Chapter 10 has produced a new book called *The Intention Experiment*.[11] Her previous book *The Field*, reviewed scientific research which showed that we are all connected together by a universal energy field. It became a bestseller and has now been translated into 14 languages. In her latest book she tackles the whole question of how human intention can be transmitted through this universal energy field and how it can effect change in our lives and in the world. This supports the spiritual concept of affirmations and suggests that collective negative thinking based on fear is responsible for creating that reality in certain parts of the world. Changing to a positive or compassionate mode of thinking could produce fundamental change for the better.

A pilot experiment has already been successful. In this, McTaggart asked a group of 16 meditators based in London to direct their thoughts to four remote targets in a laboratory in Germany. The targets were: two types of algae, a plant and a human volunteer. The meditators were asked to attempt to lower certain measurable biological variables. Significant changes were discovered in all four targets while the intentions were being sent, compared to times when the intentions were not being sent.

A further experiment was conducted at a conference in London on 10th March 2007. Here 400 people meditated and then directed their intentions to increase the biophoton emissions of a leaf in a laboratory at the University of Arizona.. The results were highly significant and could be seen on photographs taken by the sensitive imaging systems. The first Internet-based experiment was scheduled for March 24th 2007. So many people tried to log on to the website that it crashed and the experiment was rescheduled for April 14th. On that day 7,000 people participated and produced a significant increase in biophoton emissions from stringbeans.

Further work is ongoing to improve the experimental conditions by increasing the target sample sizes and making it easier to access the experiment. It seems that a concentrated group of people gathered together in the same place is more successful than a larger number of people distributed throughout the world and in contact through the Internet. This may be due to the distractions of accessing the experiment through a computer but it seems to confirm the mass-meditation experiments I mentioned earlier in Chapter 12, where large groups of advanced meditators succeeded in reducing crime and violence in society. I believe that groups together in the one place are more successful because of the interpenetration of their individual energy fields. This produces entrainment or synchronisation of individual energies, a resonance phenomenon which appears to extract energy from the vacuum to increase the power of their intentions.

The material in this book describes a technology which makes use of subtle energies and which reflects the combined effect of numerous strands of thought coming from both esoteric writings and modern scientific sources. This information is coming together at this time to enable us to create a new reality. If we do not recognise and adopt this knowledge we will be accepting a dominator society for the foreseeable future. Newgrange and the spiral serve to remind us that this present world of ours is only a shadow of what it might be. It once was a healthier and more civilised place. By understanding the meaning and purpose of Newgrange we can see how, with a knowledge of subtle energy fields, we can start rebuilding a partnership society. We can do this by the power of intention expressed from a settled state of mind.

Chapter Nineteen

Not How But Whether?

Gary Zukav in his bestselling book *The Seat of the Soul* states that we are changing from five sensory beings to multi-sensory beings.[1] As part of a general spiritual transformation that is taking place we are developing a sixth sense and coming in large numbers to accept that there is more to the world than what we see around us. However, for those who have not appreciated the sixth sense, physical survival appears to be the main basis of evolution because evolution in a spiritual sense is not detectable. Physical evolution is based on the "survival of the fittest" and therefore based on physical power and dominance. People who operate from this principle have a need to control their environment and those within it, says Zukav, and so "the basis of life in the physical arena becomes fear".

In the previous chapters I gave examples of how this sixth sense has been demonstrated as a human energy field, even though its existence is still not accepted by many scientists. As Zukav would put it, these scientists are still operating from a need to control their environment and those within it. Ideas and theories which do not fit with established knowledge can be seen as threatening the whole edifice of science and are treated with fear and suspicion, being viewed only in a five-sensory perspective.

The scientific method starts by stating a hypothesis or theory, and then trying to prove conclusively that it is either right or wrong. Scientists are not interested in results which suggest that it might be right. For a scientist it must be proved conclusively. This means that if an experiment is carried out and a certain result obtained, then others should be able to carry out the same experiment and get the same result. This is called repeatability or reproducibility.

Where scientists are dealing with results which are difficult to prove, they carry out the experiment many times and use statistics to show if the results are conclusive. For example, if a new drug is being tested on patients and only some patients experience benefits, the scientists increase the size of the sample (the number of patients) or carry out the tests repeatedly until the statistics show that there is an answer emerging which can be relied on to prove that the drug is effective or ineffective. Statistics seek to show if the result is better than would be obtained by chance and the index by which this is measured is called statistical significance.

I mentioned the Washington Demonstration in Chapter 12 where it was claimed that crime was reduced by 23%. The paper published in the peer-reviewed journal *Social Indicators Research* showed statistically that the probability that the result was due to chance was less than 2×10^{-9}, or 2 in one billion. Although this is highly significant statistically, scientists still do not recognise it, otherwise it would have hit the headlines. This shows that scientists do not accept results even when carried out according to the scientific method.

There is another factor at work here and that is the issue of credibility. If a connection cannot be made between a new or unexpected scientific result and what is already known, history shows that the result will be ignored. The problem is that scientists need to know how something works and, if they cannot see how, they tend to ignore the issue of whether.

But surely whether is much more important than how. If some new but unexplained discovery gives benefits to society can we afford to ignore it until someone comes up with a scientifically credible explanation? This is usually one based on established theories. If the explanation runs counter to established principles it usually takes a whole generation before the new theory becomes accepted. This is called a "paradigm shift" and usually requires enough time for the older scientists to retire before the new ideas can gain ground.

The direct measurement of a human energy field has fared rather badly in the past when subjected to the rigour of the scientific method. There have been countless cases reported where practitioners of dowsing, telepathy, healing or other techniques using this energy field have failed under scientific scrutiny. There are also cases where established scientists have been subjected to scurrilous attacks because their results seemed to suggest the existence of an energy field previously unknown to science. There are many reasons why the type of investigation applied in these cases may be at fault. It must be accepted that many of these techniques are delicate. The energy field needs the right environment to function.

Like the DNA I mentioned in Chapter 12, it curls up in the presence of negative thoughts. It does not seem to work in the presence of scepticism!

Jacques Benveniste worked at INSERM, a French laboratory for medical research. In his experiments he diluted antibodies in water repeatedly and found that he could still get an immune response even though so many dilutions had taken place that there could not be one molecule of the antibody left. He explained this by stating that the water had retained a 'memory' of the presence of the antibody. His research effectively proved the principle by which homeopathy works, a principle long disputed by science. His results were finally and reluctantly published in the journal *Nature* in 1988.[2] At the bottom of the article was an "editorial reservation" which stated that referees of the article commented on it with "incredulity" and that there was "no physical basis for such an activity". The editor of Nature, John Maddox, then went on to state that after publication Benveniste had agreed to allow a team of independent "investigators" to visit his laboratory and observe repetitions of the experiments.

The whole sorry story is told by one of Benveniste's fellow research scientists, Michel Schiff in *The Memory of Water: Homeopathy and the Battle of Ideas in the New Science*.[3] Schiff states that the investigation departed from any principle of objective scientific inquiry when Maddox brought in the fraud-busting magician James Randi to assist him. Having observed successful repetitions of the experiments, Schiff claims that Maddox became convinced that some type of fraud was taking place and got one of his investigators, a physicist with no experience of biological work, to prepare the microscope plates. In an atmosphere of hostility, scepticism and great pressure, the experiments were repeated and in two cases the results could not be duplicated. Maddox now claimed that these two failures negated the findings and proceeded to ignore all the successful results, even though there was a distinct possibility of error introduced by the pressure under which the work was done, the sensitivity of the experiments and the involvement of an unqualified person.

Schiff then describes how Maddox published a report in *Nature* entitled "High Dilution Experiments a Delusion". Schiff states that the report was naïve and unfair in that Maddox stated that he "was surprised to learn that experiments do not always work", a comment demonstrating his lack of knowledge of biological research and the extreme sensitivity of some of the experimental techniques. As a result of the bad publicity Schiff reports that Benveniste was ostracised and became involved in a bitter dispute with his director. This resulted in the withdrawal of his government research funding and his departure from INSERM.

In a report in *The Guardian* in March 2001,[4] it was stated that four independent research laboratories in France, Italy, Belgium and Holland have used a refinement of Benveniste's original experiment and each got a positive result. Professor Madeleine Ennis of Queen's University Belfast is quoted in the article. "Despite my reservations against the science of homeopathy" she says, "the results compel me to suspend my disbelief and to start searching for a rational explanation to our findings".

The fraud-busting magician James Randi has offered one million dollars to anyone who can prove him wrong. The BBC TV Horizon programme took up his offer and tried to duplicate Professor Ennis's results in Randi's presence. In the programme, broadcast on 26th November 2002, the results were revealed and the experiment was shown to be inconclusive. Randi kept his one million dollars. The experiments failed in the presence of extreme scepticism but were shown to work when carried out by four laboratories which were apparently neutral or certainly not hostile to the outcome.

Benveniste went on to obtain private funding and develop the memory of water idea further to include the identification of a signature for a substance in water. This signature can be transmitted electrically to imprint a test tube of distilled water. In its most simple form a test tube of biologically active substance such as ovalbumin (egg white) is placed in an electrical coil. The coil is connected to an electronic device which also has an output coil. In the output coil is placed a test tube of distilled water. The machine is switched on, and after fifteen minutes the tube in the output coil is tested by a sensitive biological test and traces of ovalbumin are detected. Benveniste claimed to be able to identify electromagnetic signals from biological molecules which should make it possible to identify toxic substances by sending their electrical signatures over telecommunications systems such as the Internet.

Apparently the idea that molecules could communicate by electrical signals is heresy to the biological establishment. In an article on his website[5] Benveniste asked why scientists are so opposed to the evolution of science. "Do these advances appear to threaten their all-too-fragile certitudes?" he asks, "such questions are not just philosophical, because these people are respected counsellors, advisers to political and industrial decision-makers…. I don't know where these mental blocks come from, but they are, in theory at least, irreconcilable with a scientist's function".

It seems that subtle energies work where there is an intention that they should or where people are unconcerned about the outcome but they do not work if there is any intention that they fail. Many people are claiming that these effects work at the quantum level. Quantum theory has shown that the observer is part of the experiment and influences the result. This is accepted now by the scientific establishment, at least by physicists but perhaps not yet by biologists. The scientific method therefore needs to be re-examined to see if the carrying out of an experiment by someone who wants it to work produces consistently different results than when the same experiment is carried out by a sceptic.

Perhaps delicate experiments need to claim 'quantum status' and not be subjected to the present level of sceptical treatment. However scientists will be reluctant to let go of the concept of 'healthy scepticism' claiming that if nobody asks the right questions we will all be taken in by scientific fraud. But what is claimed to be healthy scepticism is often a mask for political in-fighting and competition for research grants and recognition, as related by Michel Schiff in his book. His chapters include such headings as "The perverse use of legitimate technical tools", "Rumours, slanders and sarcasm", and "A psychological look at scientific repression".

The scientific establishment relies heavily on the peer review system to ensure that scientific publications are of a high standard. When a scientist carries out a piece of research and gets an interesting result, he or she will want to publish the details of the research because academics are judged by the number of their publications and by the prestige of the journals in which they appear. The paper is submitted to the editor of the appropriate journal who sends it to a panel of reviewers who are supposedly 'peers' or equals in the same field of research. In theory, good research should be recognised as such and be recommended for publication. In practice it sometimes works differently.

In one research study the peer review process was subjected to a critical evaluation. A study was carried out on two neuroscience journals to assess the extent of agreement between reviewers over a period of time. "For one journal, the relationships among the reviewers' opinions were no better than that obtained by chance. For the other journal, the relationship was only fractionally better".[6] This means that the editor would have got almost as much guidance on suitability for publication if instead of asking the reviewers he had tossed a coin in each case. Without publication in prestigious journals researchers do not get funding for new projects. The scientific establishment is thus able to subtly control who gets funding and who does not. What chance has the outsider with some new and exciting results that may turn conventional theories upside down?

We need to consider how the concept of subtle energy or torsion fields could be researched if scientists were prepared to move outside the comfort zone traditionally provided by the scientific method. This may also require the abandoning of some of the traditional academic compartmentalisation and working in multidisciplinary teams. We need a more holistic approach.

We can learn a certain amount from physics, but physics is about the physical. The energy fields I have described are beyond the normal concepts of space and time. The physiology of living things is one interface between the physical and the non-physical. Many successful trials of subtle energies have been carried out by assessing the effects of energies or energy devices on living organisms. For example, both Callahan and DeMeo give examples of tests using sprouting plants. The growth of the plants in the test can clearly be seen to be more advanced than the growth of the controls.[7] We must also begin to recognise the validity of similar subjective experiences by large numbers of people however bizarre the experiences may seem.

Metaphysics means "beyond physics" and is the branch of philosophy concerned with understanding the nature of reality. I would now like to suggest a new discipline which I have called "Experimental Metaphysics". This subject will span the disciplines of physics, biology, philosophy and psychology and will attempt to set up a new version of the scientific method which will account for the effect of human intention. Modern physics tells us that the observer is part of the experiment. Knowing what we now know about the power of intention as explained in previous chapters, how can we continue to regard sensitive measurements as repeatable, if they are carried out by different individuals with varying degrees of interest in the outcome? One approach, I suggest, is for the researcher to declare an interest at the outset of an experiment. Thus, if a researcher states that he or she is in favour of proving that an effect is true because it would benefit mankind then, if the result is positive, the experiment could be regarded as valid by those people who are prepared to apply the results with a positive intention.

I am calling this the "Qualified Scientific Method" and am proposing that it should be adopted alongside the traditional scientific method. We can no longer afford the "all or nothing" approach which has set the bar too high for new and promising technical ideas which cannot meet the criteria for repeatability, especially those concerning subtle energy and healing effects which I have described in previous chapters and which may be quantum-mechanical. The following are some tentative suggestions for procedures using the Qualified Scientific Method (QSM).

- Projects should be selected only where a number of separate pieces of information support the hypothesis. These pieces of information could come from anecdotal evidence or from publications outside the normal peer-reviewed literature.

- The researcher should declare that he or she is using the QSM and should state his or her background and the result they favour.

- The primary object of the research should be to find out if an effect is true rather than whether it conforms to established theory. A secondary object would be to find out if the effect is safe for people to use.

- Subjective reports of experiences or opinions of individuals should be acceptable if obtained in sufficient numbers. Blind tests are preferable but if they cannot be carried out, the reports of the individuals involved should be qualified by their intentions or expectations (if any).

- Statistical analysis should be used where appropriate and sample sizes should be as large as possible but studies should not be ruled out if they do not meet high levels of statistical significance.

- Attempts should be made to get articles on research using the QSM accepted by established scientific journals which could publish them in a separate QSM section. If this is not possible new QSM journals could emerge.

- Scientific research organisations and universities should set aside funds for QSM projects and should not regard this type of research as "second class".

This new Qualified Scientific Method would allow for consideration of results which had previously been dismissed according to the traditional criteria. It would be more interested in whether something is useful rather than in finding an explanation which fits with established theories. Research of this kind could investigate, confirm, dismiss or develop the many ideas and theories which I have tried to bring to public attention in this book. The results could hold the promise of enormous benefits for society.

Chapter Twenty

Creating a New Environment

In previous chapters I discussed the application of what we have learned to the individual in the form of healing the mind and the body, and I then went on to apply this to society as a whole. In this chapter I want to deal with the environment and I am using this word in its broadest sense, physical, research, energetic and spiritual. The present problems, and indeed the dominator society, arise from the perception of the individual as a separate entity having no connection or relationship with anyone or anything else. This limiting and selfish perspective arises from the identification with the ego and the body. Many spiritual writers, notably Deepak Chopra, Eckhart Tolle and Ram Dass, are showing us that we are much more than just a skin-encapsulated ego. Deepak Chopra the author of *The Seven Spiritual Laws of Success*, says that we need a sense of connectedness or an awareness of a unifying level of existence. "The ego", he says, "is not who you really are. The ego is your self image; it is your social mask; it is the role you are playing.[1]

Eckhart Tolle wrote his book *The Power of Now* [2] after experiencing a period of mental anguish. One night in his deep despair he cried out "I cannot live with myself any longer." Suddenly he stopped. "Wait a minute," he said "who is this "myself" that I cannot live with? Who is the "I" and who is the "myself?"" He realised that he was really the "I" and that the "myself " was his restless mind that was tricking him into seeing everything as separate from himself and out of his control. He saw that he was creating his own problems by his negative attitude. His mind was tricking him into dwelling in the past or the future. The answer was to be in the present, to access the Power of Now.

In his later book *A New Earth* he describes the use of the concept "I" and writes, "In normal everyday usage "I" embodies the primordial error, a misperception of who you are, an illusory sense of identity. This is the ego". He then refers to Einstein who, he says, described this illusory sense of self as "an optical illusion of consciousness".[3]

Dr. Jill Bolte Taylor, a neuroscientist at the Harvard Medical School Department of Psychiatry suffered a stroke in 1996 at the age of thirty seven. The stroke affected the left hemisphere of her brain which is concerned with reason and speech. The remarkable thing about this event is that she remained conscious throughout and was able to use her training as a neuroscientist to analyse the experience and write about it after her recovery.

In her book, *My Stroke of Insight: A Brain Scientist's Personal Journey,*[4] she relates how losing the use of the left half of her brain left her in a state of bliss where she had no awareness of her ego, her individual identity. As she recovered she realised that her ego contained all the baggage she had accumulated, including her sense of her importance as a Harvard scientist, and her conditioning by past negative events. Freed from this she saw herself as greater than the limits of her body, a universal being who was independent of events and living in a perpetual "now". Although she had temporarily lost her powers of reasoning and speech, she had gained in intuition and awareness of a greater reality. When she recovered she retained this awareness and has become a much wiser person.

As I thought about her experience, it occurred to me that the people of Newgrange must have been free from much of the baggage that we accumulate today. Living communally they must have had little need for ego and any sense of self-importance. If this was the case they must have had the awareness of a greater reality which we have lost, buried by the mental chatter of our busy lives.

Richard Alpert, later to become known as Ram Dass, was the youngest ever professor of psychology at Harvard. With Timothy Leary he started a research programme on altered states of consciousness. They found that by taking psychedelic drugs they could experience a different version of reality which they started to study and document. However, their programme fell foul of the college authorities and Alpert and Leary were asked to leave.[5]

Alpert was aware that drugs only gave a temporary 'high' and was seeking a better way to carry out his research. He found that the states of consciousness or alternative realities he and his colleagues had experienced were remarkably similar to the 'bardo' states described in the

Tibetan Book of the Dead to which he had been introduced by Aldous Huxley. At this point he decided to set out for the East and travelled through Iran and Afghanistan to India where he started looking for yogis who used meditation to reach higher states of consciousness. He eventually found a remarkable Indian saint who became his guru. Neem Karoli Baba participated in an experiment for Alpert by taking a huge dose of LSD. It did not have the slightest effect on him, thus proving to Alpert that his guru was already and permanently in a state where he was able to experience more than one reality simultaneously! Alpert decided to stay in India and study higher states of consciousness and was renamed Ram Dass or 'Servant of God' by his guru. For the last forty years he has been a spiritual teacher in the West and was the main attraction at The Prophets' Conference I mentioned at the beginning of Chapter 2.

In one taped lecture he describes his first experience of taking psilocybin, the substance in the "magic mushrooms" obtained by Timothy Leary from a Mexican shaman.[5] He had an out of body experience, objectively observing himself in his various roles and identities. He saw himself as a teacher, a cellist, a pilot and as a child. Gradually these all fell away until there was nothing there. He then looked down and he could actually see the chair underneath him. His body had disappeared! The only certainty he had left was that he was conscious. He realised that taking psilocybin had altered his consciousness so that he now accessed his fundamental state, that of pure awareness independent of any ego, identity or body.[6]

He and Timothy Leary then conducted a blind study with twenty divinity students where half of them took psilocybin and half took a placebo. Their experiences were then documented and references to taking a drug were removed. These reports were sent to religious scholars who rated them on a scale of genuine religious experience with reference to the Bible. Those that took the psilocybin scored consistently in the range of highly evolved spiritual practitioner while those who took the placebo had a low score.

This confirms what we already know, that many ancient peoples have been exploring spiritual development through mind altering drugs and it could be a basis for further research. However, when Alpert was thrown out of Harvard a remarkable opportunity for research in consciousness was also thrown out with him. Clearly, the validity of similar experiences by numbers of people resulting from properly and safely conducted research should be recognised.

I have been struck by the rejection of subjective experiences as a basis for research when these experiences relate to unusual subjects such as described above. Thus it has proved to be impossible to fund and organise larger numbers of subjects for such research projects. Yet subjective experience is used in testing drugs and in carrying out large scale research projects funded by pharmaceutical drug companies. The participants are asked to fill in a questionnaire giving their subjective experiences after taking the drugs. They are asked how did they feel? Did they experience anything unusual? It seems that experiences obtained in drug taking are valid for the purposes of curing illness but not for research in consciousness.

Other areas of research deserve to be looked at afresh by the academic community. The peer reviewed studies on the effects of advanced meditation techniques published in journals like *Social Indicators Research*, *The Journal of Conflict Resolution* and *The Journal of Crime and Justice*, should not continue to be ignored. These provide a basis for a consciousness-based solution to creating a stress-free society. They show how peace can be restored in areas of conflict, how prisons can become peaceful places and how criminals can be rehabilitated.[7] While they show how it can be done, they do not provide an explanation that fits with current scientific theory. What is apparently happening here is that the deeply settled meditative state is radiating out into society and producing a calming effect.

In terms of physics I cannot understand why Professor Myron Evans has not been given a fair hearing. He is suggesting that energy can be obtained from the so-called vacuum and I have given many examples in this book which suggest that resonance effects are drawing on unseen energy sources. Surely his research deserves some objective evaluation free from academic bias or politics. His research appears to be supported by the work of Kosyrev and others in Russia as documented by Nachalov.[8] These papers need to be translated and the work repeated in the West by unbiased researchers.

I have given many references throughout this book to different forms of subtle energies which deserve further study. These show how weather, agriculture, water and other key environmental features can be better understood and harnessed to help create a better life for all. *The Orgone Accumulator Handbook* by Dr. James deMeo is a case in point.[9] Some limited scientific studies have been carried out but these have not been accepted by peer-reviewed journals. This does not mean that they should be dismissed.

DeMeo cautions against overuse of orgone devices which can result in overcharging of orgone energy. He reports negative experiences using them in the presence of electrical appliances which seem to reverse the healing process. He also relates that the best inorganic materials are ferrous metals and that one should not try to take shortcuts by using aluminium foil as layers of inorganic material.[11] There seems to be considerable scope for a research programme to investigate the nature of this type of energy and to understand how it works and why some substances work better than others. The parallel between the organic and inorganic materials of orgone and the paramagnetic and diamagnetic materials is significant. Thankfully the days are long gone when researching this subject would leave a person liable to prosecution by the US government.

Both orgone and paramagnetic research have made use of plants as measurement devices. This appears to confirm the point made earlier that the human and also animal and plant physiologies provide an interface with subtle forms of energy. Power towers and orgone accumulators have been simply and successfully tested by surrounding them with plants or seedlings and comparing growth against a control where the energy device was absent. These are delicate forms of energy and may be affected by the electric fields of conventional measuring instruments.

We are also delicate creatures and there is a lot to be done in understanding the long-term effects of exposure to unnatural forms of energy. We have damaged the UEF around us by excessive use of radio devices and have allowed houses to be built near high voltage power lines. As mentioned earlier, a start has been made in these investigations, however, the precautionary principle, well known in science, seems to have been ignored. If you don't know the effect of something, err on the side of safety. This is the precautionary principle. Where vested interests are involved however, it seems that the principle is to increase the dose until serious illness or death occurs. This reminds me of the LD50 test I came across many years ago when getting tests done on a chemical product. Many products are tested for their level of toxicity by feeding them to rats and I was horrified to find that LD50 referred to the lethal dose at which 50% of the rats fed the substance died!

Robert O. Becker showed how small electric currents can be used to heal and even regrow bones. This indicates that we are much more sensitive to low levels of electrical energy than has previously been recognised. We should consider the effects of electromagnetic radiation as being largely unknown and reduce the levels of exposure until we can be sure that the population is safe. The safety levels for such radiation have been determined only by considering the heating effect of such radiation on the body. No consideration has been given to the effects of such radiation at the cellular level even though it has now been shown that such exposures cause cellular disruption. Dr. James L. Oschmann addresses these subjects in his book *Energy Medicine: The Scientific Basis*.[11] As well as explaining the subtle energy environment in scientific terms he also describes and gives the scientific basis for many alternative or complementary therapies.

Finally it is time to return to Newgrange and to the subject of archaeology. It is not my intention to be critical of archaeologists but I do feel that they have focussed on the wood and not seen the trees. Examining bones and pottery shards is a somewhat narrow approach and reflects the origins of archaeology as a branch of anthropology – the study of the human race. While some researchers are working on the broader social aspects of ancient peoples, archaeology has an enormous amount to contribute by considering the energies and consciousness aspects as well. This may require a multi-disciplinary approach as suggested earlier.

There is a danger that by applying our present level of knowledge to ancient sites, we may be damaging them irreparably. Are we sure that it is right to reconstruct Newgrange using modern materials? Will future generations look on the present generation of archaeologists as vandals just as we view some earlier but well-intentioned amateurs? If we come across something that we don't fully understand, is it right to dismiss it as superstition, as a primitive practice of an uncivilised people? Maybe it is

180

better to use minimally invasive techniques and if we have to reconstruct something which is falling down, to rebuild it exactly as it was in every aspect, not just its outward appearance.

My researches into the energy and purpose of Newgrange led me to see how everything is connected through the UEF. It is truly "universal" and the implications of this are that we are all connected, and that every action we perform has an effect which is felt at some level, however subtle, by every being and every piece of matter throughout the universe. Earlier peaceful societies show evidence of a highly developed intuition and a knowledge of subtle energies which we have lost. We can still see evidence of this today in some of the traditional native societies where the people live in tune with nature and cannot understand the concept of land ownership.

Modern society is living at a faster and faster pace. We are being bombarded with information and many people have come to the conclusion that it is normal to live surrounded by noise and to be constantly interrupted by phone, text or e-mail messages. The level of stress-related illness is increasing, and many people are seeking to escape by the use of alcohol or drugs. This shows that deep inside each of us is in memory that a more peaceful state exists and that we should try to find it. Unfortunately, we have been looking in the wrong place.

By understanding the connectedness of all things and all peoples we see that the pursuit of personal gain and the defence of the individual ego are strategies is which lead to ultimate disaster for society. Why should corporations constantly seek to grow larger? Why should wealthy individuals seek even more wealth? Why should we consume even more of the planet's scarce resources? The problem is that we have got onto a treadmill and we cannot get off it because we don't know how.

This book points the way on a number of fronts. It describes the dominator society and shows us clearly that we need to change. By contrasting it with the peaceful society of Newgrange, we see that there is an alternative, and that the answer is to recognize our connectedness with each other. If we do something which is harmful to another person it will ultimately be harmful for us. We also see that our health and our actions are affected by the energy field around us. On the mental level this energy field is built up from the sum of the consciousness of the individuals on the planet. If people's thoughts are based on fear or on the pursuit of their own objectives at the expense of others, this is the energy field which surrounds them and radiates out into the surrounding people and area. On the physical level we have surrounded ourselves with harmful electromagnetic and earth energies resulting from our unbridled economic growth.

We have progressed as a society from an era where humankind was dominated by religion and superstition to one in which science has taken over. But the present rigid application of the scientific method has not served us well and it is time to relax these criteria so that we can benefit from the many new ideas coming to the fore and described in this book. The new era we are entering will have a more spiritual dimension as people recognise that there is more knowledge and wisdom to be gained by going beyond that which can be detected by the five senses and by traditional scientific instruments.

My hope is that having read this book, others more qualified than I will be inspired to explore the ideas and subjects I have aired, and to progress the investigation of subtle energies in a freer and more open environment than has pertained heretofore. I have provided some further information in three appendices and the chapter notes that follow contain 220 references of which 148 refer to books and 23 refer to academic papers.

Meditation

Meditation is seen in the West as a method of stilling the mind, of relaxation and of obtaining rest. But this is just scratching the surface. In the East, meditation has always been seen as the path to enlightenment, a method of seeing a higher perspective, of contacting higher states of awareness.

In this book an attempt is made to give a scientific explanation linking the Eastern concept of higher states of awareness to the scientific concept of a unified field of energy, showing how both claim to encompass the totality of existence independent of space and time. It is claimed by some scientists that consciousness is actually the unified field since both exhibit similar properties. It has also been shown that the spiral causes matter to accelerate to enormous velocity as it approaches the centre.[1] By extension it is considered that thought, being subtle energy, is focussed and concentrated as it spirals in from the outer to the inner world through the process of meditation.[2]

There are a number of meditation techniques which have the effect of stilling the mind. These are set out below. Instructions on how to practice a simple method are given together with some advice on good meditative practice.

Concentration
This has many forms but usually involves concentrating your thoughts on your breathing or looking at a candle flame. It comes from the Buddhist tradition but no beliefs are required to practice it.

Contemplation
Here you are asked to think about a religious subject or repeat a divine name. This is usually taught in a religious setting and has a devotional context.

Guided Meditation

This is done in a group where one person leads the meditation by assisting the group in visualisation exercises. The group is usually taken to a state beyond normal experience and useful insights can be obtained.

Mantra Meditation

A mantra is literally a 'mind instrument' a word from Sanskrit which has a specific vibrational value and has been found to be effective in carrying the mind to a state of no thoughts. It comes from the Vedic tradition.

The technique of following the breath is one that is easy to learn and gives a simple introduction to meditative practice. The following extract from *The Christian's Dilemma* explains how to do it.[3]

> You should be sitting alone in a quiet room. Make sure that you are comfortable and that you are unlikely to be disturbed. Sit quietly and close your eyes. Mentally go over your body starting at your feet and moving through the whole body to your head. As you put your attention on each part of the body, make sure that you relax any tense muscles and get rid of any discomforts. If necessary take off your shoes and loosen your clothes.
>
> Once you are comfortable, start to be aware of your breath. Notice the in-breath and then follow the out-breath. Experience the rise and fall of your chest. Gradually slow down and deepen the breathing and pause briefly between breaths. Do not try to breathe too deeply.
>
> After a few minutes start counting your breaths. Count 'one' on the in-breath and then breathe out. Count two on the next in-breath etc. When you get to ten, go back to one. After a while you will get so relaxed and lazy that you won't want to count as far as ten so you can stop and go back to one after say eight, then after five, then maybe after three. Eventually you will not want to bother with counting and can just sit and 'be'.

When you want to stop, move your attention to your surroundings without opening your eyes. Become aware of sounds in the distance and then gradually let your eyes open. Take plenty of time to come out, as you will feel some discomfort if you come out too quickly. If you are disturbed in the middle of meditation, do what you have to do and go back to the meditation and then come out gradually. It is good to meditate for about twenty minutes, morning and evening, preferably on an empty stomach.

If you prefer, you can count your breath on the out breath. So you breathe in and then as you breathe out count one and so on. Meditation should never be rushed or squeezed into a time slot. You do it for the sake of doing it not for the end result. This is the practice of "mindfulness" as taught by Buddhists. I suggest that you read Thich Nhat Hahn's beautiful book *The Miracle of Mindfulness: A Manual on Meditation.* [4] In it he explains the practice of mindfulness, being fully aware of what you are doing in everything you do. He says that if you have to wash the dishes you don't do it just to be finished. You "wash the dishes to wash the dishes". Even the most boring task should be done with attention and care and not while your mind is thinking about something else.

Protection from Electromagnetic Radiation

Radiation comes from the Latin word *radius* meaning a ray. There are many kinds of radiation. Some kinds are harmful and some are not. Before we can understand how to protect ourselves from harmful radiation we must know three things, the type of radiation, its strength, and the period of exposure.

The best way to illustrate how to protect ourselves is to take an example of radiation we are familiar with – solar radiation or sunlight. Once we know that it is sunlight we know that we can screen it out by sitting in the shade, by wearing suitable clothes or by applying a skincare barrier such as suncream. Thus we can see that we can apply particular methods of screening which suit sunlight. We also know that sunlight is stronger in summer and as we travel nearer to the Equator so we can take account of the strength of sunlight by deciding where we go and when. We also know that we should not sit in the sun for too long and the weather forecast for sunny days now includes a burn time, an indication of the maximum dose of solar radiation we can receive without getting sunburnt.

Radiation is classified as ionising and non-ionising. Ionising radiation is well known to be harmful because the body is bombarded with highly energised particles which can cause cancers and genetic mutations. The higher the frequency the higher the energy and the more dangerous, as can be seen in the table below. So we know that we have to avoid nuclear radiation and only receive very small doses of X-rays at a time. As we move down the frequency spectrum we come to light waves and radio waves which together are called electromagnetic radiation and which contain less energy than ionising radiation.

Electromagnetic (EM) radiation covers a wide spectrum. It ranges from low frequency electricity in the home at the lower end to light waves and gamma rays at the upper end. The table gives a summary of the electromagnetic spectrum. The basic unit of frequency is the Hertz = 1 cycle per second. Kilohertz = kHz is 1000 Hz (10^3). Megahertz is 1 million Hertz (10^6). Gigahertz is 10^9 and Terahertz is 10^{12} Hertz.

Type of Radiation	Frequency	Advice on exposure
Ionising radiations		
X-rays, gamma rays	Very high – highly energised particles	Avoid at all costs
EM (non-ionising)		
Ultraviolet	Above 800THz	Avoid if possible (Causes sunburn, used for tanning)
Visible light	480-700THz	Reduce exposure to direct strong sunlight
Near infrared	300THz	Used for night vision and heating. Any overexposure will usually be felt as heat.
Microwaves	3Ghz -30GHz	Minimise exposure to radiation from microwave ovens especially if old or in poor condition
Mobile phone band (subset of UHF band)	900MHz – 1800Mhz	Avoid base station aerials, WiFi hot-spots and overuse of mobile phone
UHF TV band	300MHz -3GHz	Choose to live away from transmitting stations
VHF/FM radio and TV bands	80-300MHz	Choose to live away from transmitting stations
Radio long/medium wave	300kHz – 3000kHz	Safe if not too close to transmitter
Mains electricity	50Hz, 60Hz(US)	Avoid high voltage lines, substations, large transformers and motors

EM radiation is non-ionising and therefore not as harmful as ionising radiation and for this reason we have been lulled into a false sense of security. However, in recent years both the quantity and quality of EM radiation has changed dramatically. No research has been done considering the combined effects of a number of radiation sources acting together. It is now wise for us to consider all the points made above concerning sunlight and apply them to EM radiation if we want to protect our health. We should consider the possibility of screening and investigate the type of screening needed as it varies for different types of EM radiation. We should also know the strength of the radiation and how long we are exposed to it. As well as this we should also consider the cumulative effect of being exposed simultaneously to numerous forms of EM radiation from power lines, mobile phone and other radio transmitters and from the wide range of modern electrical equipment.

As the frequency increases, the ability of waves to follow the curvature of the earth decreases. Low and medium frequency waves are able to travel further because they are reflected off the ionosphere, a layer of ionised particles in the upper atmosphere. Higher frequency waves cannot do this and the range of FM radio stations is limited to about 50 miles and mobile phone masts have an even shorter range. They are dotted around the landscape to get adequate coverage. There are over 30,000 mobile phone base stations in the UK and 43% of the population live within 500 metres of one. Readers in the UK can find out how near they are to a base station at https://mastdata.com/ and in the Republic of Ireland at https://coveragemap.comreg.ie/map

Screening
Electric fields can be screened by interposing electrically conducting materials such as metals. Ideally one should attempt to create a Faraday cage. This is used in laboratories where it is necessary to avoid all electrical interference and is an area completely enclosed by metal. This is not practical in the home and the best that can be obtained is partial screening

Mains electricity will generally only emit low intensity electric fields except where one lives close to a high voltage power line. Otherwise, the main concern in the home is for screening from mobile phone masts if one is close by. This can be done by painting walls with nickel paint, covering windows and beds with metal mesh curtains and using aluminium foil. Any screening metal used should be earthed as otherwise it could act as an aerial and actually increase the signal strength in places. Dr. DeMeo cautions against the use of aluminium foil.[1] DeMeo has also published an interesting report on the effectiveness of orgone blankets in screening out electric fields.[2]

It is almost impossible to screen out magnetic fields and therefore people should avoid living near electricity substations if possible. Strong magnetic fields are also emitted by powerful electric motors.

Strength of Radiation
Radiation is stronger nearest to the source and falls off with the square of the distance from it so you are four times as safe at a distance of 200 meters as you are at a distance of 100 meters Thus, most signals are not very strong unless you are very close to a transmitter. The general rule is to locate yourself as far away as is practical from sources of radiation..

Mobile phones emit stronger signals if the base station is far away or if used in a car which is like a partial Faraday cage. Avoid prolonged use of a mobile phone in weak signal areas and in a car. Stop the car, get out and stand beside it if you have to make a call. Another way to reduce exposure is to use a hands-free kit which reduces exposure to the head. As well as reducing the risks of radiation this also makes for safer driving.

There has been a lot of misinformation about mobile phones. The first point to note is that mobile phones work in cells which are quite small. In a city a cell might only cover a square mile so as you travel, you are automatically transferred from an aerial in one cell to an aerial in the next one. Because the cells are small the power needed is also small. Typically a mobile phone has a power output of 2 watts, about the same as a battery operated torch, and a base station has a power output of about 50 watts, about the same as a light bulb.[3]

Exposure
Many problems only arise due to prolonged exposure such as may happen if you spend a long time in the same place each day such as at work or sleeping. It is wise to avoid using old computers with cathode ray tubes. These are the ones which look like TV sets. Buy an LCD screen. Do not

sit close to an old television set. Do not use electric blankets which are switched on while you sleep. Do not sit near a microwave oven in the kitchen while it is on. You can also reduce the amount of unwanted radiation by switching off equipment and not leaving it plugged in on standby. If it shows a small signal light or indicator, switch it off or plug it out. This will also reduce your electricity bill and help reduce carbon emissions.

What level is safe?
We cannot say what level is safe because we do not know for sure what the long term effects of EM fields are. Most transmitters work below the maximum radiation level permitted by the standards but there is little agreement on what these standards should be. They are mainly based on the possible heating effects of EM radiation but little is known about the long term effects at the cellular level. We also know that some people are more susceptible to EM fields than others so research carried out on the general population may not be appropriate to these people.

On May 23rd 2007 the BBC TV Panorama team broadcast a programme concerning the rapid increase in WiFi, the wireless system that enables laptop computers to access the Internet.[4] They showed that the level of radio frequency signals from WiFi systems was up to three times the intensity of the main beam from a mobile phone mast. These systems are being widely installed in schools even though the British government has recommended that mobile phone masts be sited away from schools and playgrounds.

More worrying still was an interview broadcast with Sir William Stewart, Chairman of the Health Protection Agency which had advised the government on mobile phone emissions. He stated that there may be possible biological effects from these signals and that a more precautionary approach should be adopted. A number of scientists from other countries were interviewed and expressed concern that evidence of biological damage was being ignored. Dr Gerd Oberfeld, from Salzburg called for Wi-Fi to be removed from schools. He said: "If you go into the data you can see a very very clear picture - it is like a puzzle and everything fits together from DNA break ups to the animal studies and up to the epidemiological evidence; that shows, for example, increased symptoms as well as increased cancer rates." If you see a sign that says

"WiFi Hot-Spot" avoid that area. These hot-spots are becoming common in most cities and are part of what is increasingly being termed "electrosmog".

Resources

A very good source of information is Powerwatch an organisation giving information, advice and products for EM protection. Their website www.powerwatch.org.uk has useful links to many other sites.

Working with Subtle Energies

Raise your hands in front of your face with palms facing each other. Now bring your palms closer together slowly with your eyes closed. You should begin to feel some resistance or heat when your hands are about an inch (2-3cm.) apart. If you feel this stop and open your eyes. How close are your hands to each other? Did you feel some pressure or heat on your palms? What you have felt is the spiral energy of healing which we all have in our palms.

The next exercise is the one which is at the beginning of the book.

Point the index fingers of each hand towards each other and bring them together so that the tips of your fingers are about a third of an inch apart (1 cm.). Do this against a plain but dark background. You should see a thin light grey or violet stream of energy going between your fingers. What you have done is to focus the energy of your aura onto a point (your fingertips) where it becomes easier to see. Now move your fingers to and fro relative to each other and watch the energy stream follow the movement.

These are some simple examples of subtle energy fields. Some fields involve energy and some involve information. We can also sense energy fields by techniques such as dowsing which is a bit of both, energy and information. You can also do other simple experiments which are about collecting information using telepathy and another technique called remote viewing.

Dowsing
Dowsing is well known and is used for finding water or minerals under the ground. Although it is not accepted by science it is widely used by industry and by farmers who are more interested in results than theory. Clearly there are some people who are more gifted than others but everybody has some ability and I was surprised to find that I could do it and now I use it regularly especially in making decisions.

Dowsing is also called divining but divination is a wider subject and could include getting information from other sources such as astrology or Tarot cards. Broadly speaking we can break dowsing down into two activities, looking for things and looking for information. Looking for things is usually done with dowsing rods and looking for information is done with a pendulum.

We will concentrate on the pendulum since it is easier to make one and you can try it straight away. Any small object can be attached to a piece of thread or string and will act as a pendulum. Tie it onto the thread and hold the thread so that there is about eight inches (20cm.) between your finger and the object. Most people experience the answers given by the pendulum as a rotation, clockwise or anticlockwise. One direction is yes and the other is no and the first thing to do is to calibrate the pendulum, to find out which is which.

Holding the pendulum thread give the thread a slight push so that the pendulum gently swings two and fro towards and away from you. Ask a question to which you know the answer is yes. For example, "is my name (your name)?" You should experience some gentle rotation in one direction. Now ask, "is my name (another name)?" You should get a movement in the opposite direction. Now you know which is which. For me yes is always anticlockwise. For other people it may be the reverse.

Your questions should be carefully thought out, logical and unambiguous. They should be phrased in a way which can be answered by a yes or a no.

I have found the following tips to be helpful:

1. It is better to ask a question where you genuinely do not know the answer or do not have strong preferences. "Should I take an umbrella with me today?" is a good example. "Does X have cancer?" is not a good example as you may have a strong preference for a no.

2. Your motives should be the highest good. For example, it does not seem to work if you use it to ask if a certain horse is going to win a race, or anything to do with gaining at the expense of others.

3. You must be absolutely clear in the question you ask. If you ask, "Should I go to the meeting or not?" you will not get useful answer as your question did not have a clear yes/no structure. Just ask, "Should I go to the meeting?"

The pendulum can only answer questions which have a yes/no answer but you can learn by carefully constructing your questions to use the pendulum to get a lot of information. For example if you wish to know your birth time for an astrology reading you can ask, "Was I born between midnight and noon?". If the answer is yes you can ask, "Was I born between midnight and 6am?" If the answer to the first question was no you can ask, "Was I born between noon and 6pm?" This way you progressively narrow down the time until you get a sufficiently useful answer.

Dowsing with rods is more complex but you can study it from books or on the Internet (see below). You can use the rods to check for harmful energies in your house, to locate pipes or wires, to find water or other hidden objects. You can buy dowsing pendulums and you can make dowsing rods from wire coat hangers.

Resources
Irish Society of Diviners: www.irishdiviners.com/
British Society of Dowsers: www.britishdowsers.org/

Telepathy
Telepathy is a kind of extrasensory perception and comes from the Greek *tele* meaning distant and *pathe* meaning feeling. Rupert Sheldrake and his colleagues have carried out extensive research on the subject and this is described in his book *The Sense of Being Stared at and Other Aspects of Extended Mind.*[1] The book also describes numerous experiments you can do yourself.

Most people have experienced some form of telepathy especially with people who are close to them. You may say something and your partner says, "How did you know I was thinking about that?" Recently my wife, Lynda and I were finishing a walk in the Wicklow mountains. It was a hot day and I was thirsty. I was thinking, "Will I have beer or cider when we get to the pub?" Lynda asked me out of the blue, "Are you going to have cider when we get to the pub?"

The phone rings and you know who it is before they speak. You are staring at somebody and they turn around and look at you. If you have a dog you may notice that it gets up and sits at the door ten minutes before its owner comes home, regardless of the time.

Try keeping a notebook beside your phone. When the phone rings as you walk towards it ask yourself who is phoning. Write down in the book who phoned and who you thought it was. After a week or so add up the number of times you were right. The chances of being right depend to some extent on the pattern of your calls. If you have a friend who phones regularly you may be just guessing correctly but if your calls are from a variety of people, friends and strangers, then a high score would suggest some telepathy. In general the research suggests that telephone telepathy is stronger between people who are in close relationships.

I have occasionally tried staring at people across the road or from my car. I know this is bad manners but I have been amazed at the number of times the person turned and looked at me and I had to look away. It seems that staring at people even through closed circuit TV can produce the same reaction. In controlled experiments, stress responses were measured in participants which correlated with the times at which they were being stared at. [2]

Remote Viewing
Many people have demonstrated the ability to see things which are remote from them, hence the name for this activity. Research into remote viewing was funded by the US government for thirteen years. They used this technique to find out what was happening at locations in the Soviet Union and elsewhere during the Cold War.

I had an interesting remote viewing experience which I described in *The Christian's Dilemma:*

When my son Mark was a teenager he used to go for long cycles on his racing bike. One day when he had not come back by the expected time, my wife and I started to get worried. I had been reading a book about remote viewing and decided to give it a try. Remote viewing is a technique by which ordinary people are able to see things which are distant from them. I sat down and closed my eyes. After I had spent some time clearing my mind and relaxing, I expressed to myself a desire to know where Mark was. I then saw a picture of him walking up a hill pushing his bicycle. The place looked like South Avenue, a road about a quarter of a mile from our house. This was puzzling because I would have expected him to come home by a different route. He arrived about five minutes later saying that his tyre had punctured and he had walked for miles. I asked him what route he had taken and he explained that he had come by South Avenue because it was safer to walk that way than on the main road.[3]

The history of remote viewing is interesting. Hal Puthoff was a scientist working at the Stanford Research Institute in California. The CIA had learned that the Russians were doing psychic research to spy on the US so they decided that the US should also do the same. When the CIA visited research labs they were laughed at because no scientists would take extrasensory perception or clairvoyance seriously.[4] However Puthoff had already been working with Ingo Swann, a gifted psychic, and was able to take on the project which in collaboration with a colleague Russell Targ, ran for thirteen years.

They recruited a team of skilled clairvoyant people who were able to perfect their skills and identified the details of distant locations right down to the descriptions of what was going on inside high security installations. Numerous examples from this work are given in *Miracles of Mind* by Russell Targ and Jane Katra.[5] Some of the results were highly significant statistically and were published in the Proceedings of the Institute of Electrical and Electronic Engineering (IEEE).[6]

One test attempted to see if long distances affected the results. Ingo Swann agreed to view the planet Jupiter just before the Pioneer 10 space probe flew past. Swann said he saw a ring around Jupiter and thought that he had got confused with Saturn. But later NASA reported that Jupiter had a ring around it at that time.[7]

Remote viewing is quite easy to test for yourself. Arrange with a friend who is travelling to note where they are and their surroundings at fixed times. At those times sit down in a quiet place and gently see if you can get a picture of where your friend is. Make notes and do sketches. Compare notes when your friend returns. The skill is to learn to separate out your preconceived ideas from the actual message. It is better not to concentrate too hard.

Chapter 1   The Mysteries of Neolithic Ireland

1. The Illustrated Guide to the Megalithic Cemetery of Carrowmore Co. Sligo, Gŏran Burenhult (author and publisher), Revised Edition 2001.
2. Ibid. page 6.
3. Ibid. page 5.
4. *Newgrange and the Bend of the Boyne,* Sean P O'Riordain and Glyn Daniel, London 1964, page 19.

Chapter 2   From Partnership to Domination

1. *The Chalice & the Blade: Our History, Our Future*, Riane Eisler, Harper Collins, San Francisco, 1995.
2. Michael Grant states that Constantine was an autocrat who 'believed that he could kill anyone'. See *Constantine the Great: The Man and His Times*, Michael Grant Schribner & Sons New York, 1994 See also Eisler, *The Chalice & the Blade,* page 131.
3. *The Gnostic Gospels*, Elaine Pagels, Penguin, London, 1990, page 17.
4. *The Christian's Dilemma: A guide to the new spirituality*, Kieran Comerford, Elo Publications, Dublin, 2002, page 18. Also *Reincarnation: The missing link in Christianity,* Elizabeth Clare Prophet, Summit University Press, Corwin Springs, Montana, 1997, pages 221-2.
5. *The Catholic Encyclopaedia,* Robert Appleton & Co. New York 1907-14, s.v. 'Origen' as quoted in Prophet, *Reincarnation: The missing link in Christianity* page 223.
6. Prophet, *Reincarnation: The missing link in Christianity*, page 244.
7. *The Holy Blood and the Holy Grail,* Michael Baigent et al, Corgi Books, London 1983.
8. *Renewing the Irish Church,* Joe McVeigh, Mercier Press, Cork 1993, page 28.

9. *The Chalice & the Blade: Our History, Our Future*, Riane Eisler, Harper Collins, San Francisco, 1995.
10. Eisler, *The Chalice & the Blade,* pages 29- 30.
11. Ibid., page 36., also *The Minoan World*, Arthur Cotterel, Michael Joseph, London 1979 pages 10, 123.
12. Eisler op. cit., pages 32-33
13. BBC Timewatch programme broadcast on 20th April 2007. See Lilley, H., http://news.bbc.co.uk/2/hi/science/nature/6568053.stm
14. See www.fmnh.org
15. Broadcast on 31st January 2002, See www.bbc.co.uk/science/horizon/2001/caraltrans.shtml
16. *What is Céide Fields?* Interpretive centre brochure by Dr. Seamus Caulfield, 1992.
17. Ibid.
18. *Keeper of Genesis: A Quest for the Hidden Legacy of Mankind*, Robert Bauval and Graham Hancock, Arrow Books, London 1997, page 27

Chapter 3   The Spiral

1. The Heritage Awareness Group (HAG) has become part of the Irish Druidschool www.druidschool.com
2. *The Boyne Valley Vision,* Martin Brennan, The Dolmen Press, Portlaoise 1980 page 16.
3. *Irish Symbols of 3500BC,* N L Thomas, Mercier Press, Cork, 1988, also *The Stars and the Stones,* Martin Brennan, Thames and Hudson, London 1983.
4. *Newgrange: The Mystery of the Chequered Lights,* Hugh Kearns, New Island, Dublin 2005.
5. Ibid. page 12.
6. This idea was suggested to me by Michael Roberts of Sligo.
7. *Stone Age Soundtracks: The Acoustic Archaeology of Ancient Sites*, Paul Devereux, Vega, London, 2001.
8. Ibid. page 101.
9. I am indebted to Joe Dunne of Dublin for this insight.
10. *Exploring Newgrange*, Liam Mac Uistin, The O'Brien Press, Dublin, 1999, page 71.

11. Dronfield, Jeremy, 1996, "Entering Alternative Realities: Cognition, Art and Architecture in Irish Passage-Tombs" *Cambridge Archaeological Journal*, 6(1):37-72.
12. *The Vortex of Life,* Lawrence Edwards, Floris Books, 1993.
13. *The Mystic Spiral*, Jill Purce, Thames & Hudson, London 1974.
14. See *Introduction to Sanskrit,* Thomas Eugenes, Point Loma Publications, San Diego 1989. The phrase 'collapse of the fullness' comes from the word 'akshara' meaning literally a consonant or that which ends the full sound of a vowel. A deeper meaning is the 'kshara' of A, the collapse of the inherent pulsation of the universe, the collapse of infinity to the point, or the collapse of the wave function in quantum mechanics, the mechanism by which consciousness creates its own reality. See *The Neurophysiology of Enlightenment*, Robert Keith Wallace, MUM Press, Fairfeld, Iowa 1986, page 7.
15. Wallace, pages 9-23.
16. *Irish Prehistory: a social perspective*, G Cooney and E Grogan, Wordwell Dublin, 1994, pages 55,57.
17. Ibid. page 54.
18. This is based on my own observations, discussions with others and the published statements that the builders of Newgrange were wealthy farmers. See *Newgrange Archaeology, Art and Legend,* Michael J O'Kelly, Thames & Hudson, London 1982, Cooney and Grogan above, and *Knowth and the Passage Tombs of Ireland,* George Eogan, Thames & Hudson, London 1986.
19. Yoga Sutras of Patanjali 2.35. See *Effortless Being: The Yoga Sutras of Patanjali*, translated by Alastair Shearer, Wildwood House, London 1982.

Chapter 4   Torsion Fields

1. *Living Energies,* Victor Schauberger, Callum Coats (Editor), Gill & Macmillan, Dublin 2001
2. This results from the principle of the conservation of angular momentum. The angular momentum of a body is its moment of inertia multiplied by its angular velocity. Since angular momentum is proportional to the radius, halving the radius must double the angular velocity if the angular momentum is to be

conserved. This means that as the radius approaches zero the angular velocity must approach infinity!

3. *The Coming Energy Revolution*, Jeane Manning, Avery Publishing Group, New York, 1996.
4. *A White Paper On The Law of Cause and Effect,* William A. Tiller, www.tiller.org
5. *The Field,* Lynne McTaggart, Element, London 2001, pages 41-2.
6. *Wholeness and the Implicate Order*, David Bohm, Routledge & Kegan Paul London 1980.
7. See http://www.amasci.com/freenrg/tors/tors3.html and www.rexresearch.com/torsion/torsion1.htm
8. See www.coralcastle.com
9. Ibid. See also *The Giza Power Plant: Technologies of Ancient Egypt* Christopher Dunn, Bear & Company Inc., Santa Fe, New Mexico, 1998, page 117.
10. See www.keelynet.com/davidson/npap1.htm
11. PCT patent specification WO 86/05852
12. See http://www.amasci.com/freenrg/tors/tors3.html

Chapter 5   Overcoming Gravity

1. *Exploring the Physics of the Unknown Universe*, Milo Wolff, Technotran Press, California, 1990. The equation for the force of gravity was discovered by Newton. $F = Gm_1m_2/r^2$ where m_1 and m_2 are two masses and r is the distance between them. G is the gravitational constant 6.6259×10^{-11} m^3/kg-sec^2. Thus it can be seen that the force decays as the square of the distance and since the gravitational constant is very small, the force of attraction between two everyday objects of similar size is tiny, for example, the gravitational force between two bowling balls each weighing one kilogram at a distance of one metre apart would be 6.6259×10^{-11} kilogram metres per square second.
2. See http://www.aias.us/
3. See http://www.aias.us/documents/eceArticle/ECE-Article_EN.pdf Section 6.
4. *The Giza Power Plant: Technologies of Ancient Egypt* Christopher Dunn, Bear & Company Inc., Santa Fe, New Mexico, 1998, pages 6-12.

5. Ibid. Chapters 1-4.
6. *The Bridge to Infinity,* Bruce L. Cathie, Adventures Unlimited, Illinois, 1997. See also: http://www.bibliotecapleyades.net/ciencia/antigravityworldgrid/ciencia_antigravityworldgrid08.htm
7. See http://morris108.wordpress.com/2008/06/11/pictures-sufis-really-levitating-an-80-kilo-stone-india/
8. Yoga Sutras of Patanjali 3.42. See *Effortless Being: The Yoga Sutras of Patanjali*, translated by Alastair Shearer, Wildwood House, London 1982.

Chapter 6   Ancient Connections

1. *When the Sky Fell: In Search of Atlantis*, Rand and Rose Flem-Ath, Orion Books, London 1996.
2. Ibid. page 46.
3. *Gateway to Atlantis: The Search for the Source of a Lost Civilisation*, Andrew Collins, Headline Book Publishing, London 2000, page 13.
4. *The Chalice & the Blade: Our History, Our Future*, Riane Eisler, Harper Collins, San Francisco, 1995. page 63
5. Collins op. cit.
6. *Keeper of Genesis: A Quest for the Hidden Legacy of Mankind*, Robert Bauval and Graham Hancock, Arrow Books, London 1997 page 16 et seq.
7. Ibid. page 9
8. Collins op. cit., page 67
9. *From Carnac to Callanish: The prehistoric stone rows and avenues of Britain, Ireland and Brtiitany*, Aubrey Burl, Yale University 1993
10. *New Grange and other incised tumuli in Ireland*, George Coffey, Dolphin Press, Dorset 1977. (first published in 1912)

Chapter 7   Layers Within Layers

1. *Saharasia: The 4000BCE Origins of Child Abuse, Sex-Repression, Warfare and Social Violence in the deserts of the Old World,*

Natural Energy Works 1991, summary at: http://www.orgonelab.org/saharasia.htm

2. See http://www.scienceagogo.com/news/culture_weather.shtml and Brooks, N. 2004. Beyond collapse: the role of climatic desiccation in the emergence of complex societies in the middle Holocene. In Leroy, S. and Costa, P. (Eds.) *Environmental Catastrophes in Mauritania, the Desert and the Coast. Abstract Volume and Field Guide. Mauritania,* 4-18 January 2004. First Joint Meeting of ICSU Dark Nature and IGCP 490. See http://www.cru.uea.ac.uk/%7Ee118/publications/publications-subj.html

3. See http://www.orgonelab.org/suppression.htm

4. See www.orgone.org the website of PORE public Orgone Research Exchange.

5. See http://www.orgonelab.org/funding.htm

6. *Newgrange Archaeology, Art and Legend,* Michael J O'Kelly, Thames & Hudson, London 1982, page 85.

7. I am indebted to Con Connor of the Heritage Awareness Group who first introduced me to the work of Wilhelm Reich by giving me a photocopy of one of the 'burnt' books on orgone accumulators and for drawing my attention to the possibility that Newgrange was designed as an orgone accumulator. This is also mentioned at http://www.orgone.org/articles/ax6antot.htm under the heading 'Energy focusing devices'.

8. Personal discussion with Jan DeVries. In the context of human energy and creating new realities it is worth reading *Body Energy* (1989) and *The Miracle of Life*, (1997), Jan DeVries, Mainstream Publishing Co. Edinburgh.

Chapter 8   Stages of Development

1. *Sun Moon and Standing Stones*, John Edwin Wood, Oxford University Press, 1978.

2. Ibid. page 81.

3. *Knowth and the Passage Tombs of Ireland,* George Eogan, Thames & Hudson, London 1986.

4. *Ancient Ireland: Life Before the Celts*, Laurence Flanagan, Gill & Macmillan, Dublin 1998, page 66-7.

5. Eogan op.cit. Chapter 3
6. *Earth Magic*, Francis Hitching, Cassell, London 1976.
7. Ibid. page 121.
8. Although a work of fiction, Bruce Chatwin's book *The Songlines*, Penguin 1988, describes them very well.
9. *Newgrange Archaeology, Art and Legend,* Michael J O'Kelly, Thames & Hudson, London 1982, page 128.

Chapter 9   Newgrange is Damaged

1. *Newgrange Archaeology, Art and Legend,* Michael J O'Kelly, Thames & Hudson, London 1982, 2004 edition, pages 109-114.
2. *Newgrange and the Bend of the Boyne,* Geraldine Stout, Cork University Press, 2002, Figure 13, page 44.
3. *Newgrange,* Geraldine Stout and Matthew Stout, Cork University Press, 2008, pages 4-6.
4. See http://homepage.eircom.net/~seanjmurphy/irhismys/newgrknow.htm This website includes details of the concrete construction blocking the eastern entrance to Knowth as reported in the Sunday times, Irish Edition on 22nd October 2000.

Chapter 10   The Lessons of Newgrange

1. *Tuning the Diamonds*, Susan Joy Rennison, Joyfire Publishing, Staffordshire 2006. Pages 102-108.
2. *The Value of Science,* a public address delivered by Dr. Richard Feynman to the autumn 1955 meeting of the National Academy of Sciences and quoted in *What Do You Care What Other People Think,* Richard P. Feynman as told to Ralph Leighton, Unwin Paperbacks, London, 1988 pages 247-248.
3. *The Orgone Accumulator Handbook*, James DeMeo, Natural Energy Works, Ashland, OR. 2nd ed. 1999.

Chapter 11   The Revival of Shamanism

1. See www.shamanismireland.com
2. *The Holotropic Mind*, Stanislav Grof, Harper Collins, San Francisco 1993, page 15
3. Ibid. page 18
4. Ibid. page 84
5. Rig Veda 1.164.39. See Chapter 2 with reference to the spiral.
6. *The Way of the Shaman*, Michael Harner, Harper and Rowe New York 1980.
7. *The Journey to You: A Shaman's Path to Empowerment,* Ross Heaven, Bantam, London 2001
8. *Experiencing Ritual: A New Interpretation of African Healing*, Turner, E.. Philadelphia: University of Pennsylvania Press, 1992
9. http://www.shamanism.org/articles/article02.html
10. Heaven, op. cit. page 119
11. *Irish Symbols of 3500BC,* N L Thomas, Mercier Press, Cork, 1988, also *The Stars and the Stones,* Martin Brennan, Thames and Hudson, London 1983
12. *Sacred Waters: Holy Wells and Water Lore in Britain and Ireland,* Janet and Colin Bond, Granada, London, 1985, pages 2 and 3
13. *Newgrange: The Mystery of the Chequered Lights,* Hugh Kearns, New Island, Dublin 2005.
14. *The Secret Life of Water*, Masaru Emoto, Simon & Schuster, London 2005.
15. *Stressed out? Put on a blindfold for 72 hours and bang into chairs*, The Observer June 5, 2005. See also http://www.sacredtrust.org/
16. Eriksen, P. (2008), The Great Mound Of Newgrange. *Acta Archaeologica,* 79 pages 250–273.

Chapter 12   The Universal Energy Field

1. *The Holographic Universe,* Michael Talbot, HarperCollins, London 1996, page 13.
2. *The Field,* Lynne McTaggart, Element, London 2001, pages 124-5.

3. *Vibrational Medicine for the 21st Century*, Richard Gerber, Piatkus, London 2000.
4. McTaggart op. cit. page 153.
5. Ibid page 176-7.
6. See http://noosphere.princeton.edu.
7. Ibid.
8. Wallace R K, Physiological Effects of Transcendental Meditation, *Science* 167, (1970) pages 1751-1754.
9. Orme-Johnson D W et al, 'Intersubject EEG Coherence: is consciousness a field?' *International Journal of Neuroscience* 16, (1982) pages 203-209.
10. This event was witnessed by Luke Chan author *101 Miracles of Natural Healing* and details can be obtained at www.chilel-qigong.com
11. Poponin V, 'The DNA phantom effect: direct measurement of a new field in the vacuum substructure' *Journal of Nanobiology*, 1998 cited in *The Isaiah Effect: Decoding the Lost Science of Prayer and Prophecy*, Gregg Braden, Three Rivers Press, New York, 2000, pages 207-210. Also "Modulation of DNA Conformation by Heart Focused Intention" McCraty et al, 2003, Institute of HeartMath, www.heartmath.org
12. Hagelin John S. et al, 'Effects of Group Practice of the Transcendental Meditation Program on Preventing Violent Crime in Washington, D.C.: Results of the National Demonstration Project, June-July 1993' *Social Indicators Research* 47, no. 2 (June 1999): 153-201.
13. Orme-Johnson D W et al, 'International Peace Project in the Middle East' *Journal of Conflict Resolution* 32, no. 4 (December 1988 pages 776-812).
14. *The Field,* Lynne McTaggart, Element, London 2001, pages 276-7.

Chapter 13   As Above, So Below

1. Keeper of Genesis*: A Quest for the Hidden Legacy of Mankind*, Robert Bauval and Graham Hancock, Arrow Books, London 1997 page 146 et se.

2. *The Orion Mystery* Robert Bauval and Adrian Gilbert, William Heinemann Ltd., London 1994.
3. Bauval and Hancock op. cit., chapters 3 and 4
4. *The Giza Power Plant: Technologies of Ancient Egypt* Christopher Dunn, Bear & Company Inc., Santa Fe, New Mexico, 1998, pages 6-12.
5. Dunn op. cit. page 145.
6. See www.mikepettigrew.com/earthchanges
7. *Light on Life: an introduction to the astrology of India*, Hart Defouw and Robert Svoboda, Arkana, London 1996.
8. *Calendars and Constellations of the Ancient World*, Emmeline Plunket, Senate 1997, first published 1903 pages 109 et seq.
9. *The Circle of Stars: an introduction to Indian astrology,* Valerie J. Roebuck, Element 1992.
10. *Living the Field: Earth Energies,* Lynne McTaggart, http://www.theintentionexperiment.com/learning-aids
11. Ibid. Pages 33-39.
12. *A New Science of Life: The Hypothesis of Formative Causation,* Rupert Sheldrake, Blond and Briggs, London, 1981
13. *The Rebirth of Nature; The Greening of Science and God,* Rupert Sheldrake, Park Street Press, Vermont 1994 page 111.
14. ibid. page 169.
15. Ibid. page 176.
16. *Dialogues with Scientists and Sages; The Search for Unity,* Renée Weber, Routledge & Kegan Paul, London 1986.
17. ibid. page 86.

Chapter 14   The Energy Field is Damaged

1. *The Biology Of Transcendence,* Joseph Chilton Pearce, Park Street Press Vermont, 2002. See also www.heartmath.org.
2. *The Body Electric,* Robert O. Becker and Gary Selden, Quill, New York, 1985.
3. *Cross Currents, The Perils of Electropollution, The Promise of Electromedicine,* Robert O. Becker, Jeremy P. Tarcher/Penguin, 1990.
4. See www.energyfields.org/science/becker.html
5. The Times, London, 3rd June 2005

6. *The Vortex of Life,* Lawrence Edwards, Floris Books, 1993. pages 225-226.
7. *Safe as Houses*, David R. Cowan and Rodney Girdlestone, out of print but available on line at www.leyman.demon.co.uk/index.html
8. See http://www.cogreslab.co.uk/sick_build.asp
9. *Vastu: The Origin of Feng Shui,* Schmieke Marcus, Goloka Books Ltd., 2002. See also http://www.veda-academy.com/departments/vasati.htm
10. *The Feng Shui Handbook*, Master Lam Kam Chuen, Gaia Books, London, 1995.
11. See http://www.feb.se/index_int.htm

Chapter 15   Healing the Energy Field

1. *Safe as Houses*, David R. Cowan and Rodney Girdlestone, out of print but available on line at www.leyman.demon.co.uk/index.html
2. *Pi in the Sky*, Michael Poynder Rider, London 1992
3. *Newgrange and the Bend of the Boyne,* Geraldine Stout, Cork University Press, 2002, page 197.
4. Poynder. page 41
5. See http://www.britishdowsers.org/EEG_site/archive/articles/arc2004_issue34/ GreatDowsers_Part3.htm
6. See http://www.pamex.com/reports/coghillreport.pdf, also see http://www.landandspirit.net/html/geopathic_stress.html This contains a very informative 45 page article on geopathic stress. See, in particular, the case histories on the position of beds and on earth acupuncture.
7. See http://www.who.int/peh-emf/standards/EMF_standards_framework%5B1%5D.pdf
8. *Cross Currents, The Perils of Electropollution, The Promise of Electromedicine,* Robert O. Becker, Jeremy P. Tarcher/Penguin, 1990, pages 269-276
9. See www.powerwatch.org.uk
10. See www.geomantica.com

11. *Chainsaw Massacre*, Lucy Siegle, Observer Magazine, 12th October 2008.

Chapter 16   Magnetic Attraction

1. See http://www.keelynet.com/unclass/magcurnt.txt
2. See http://www.aias.us/documents/eceArticle/ECE-Article_EN.pdf   Section 6.
3. See http://www.cheniere.org/ also Bearden et al., "Explanation Of The Motionless Electromagnetic Generator With 0(3) Electrodynamics", *Foundations of Physics Letters, Vol. 14., No. 1, 2001*
4. Y. Aharonov and D. Bohm, "Significance of Electromagnetic Potentials in the Quantum Theory," *The Physical Review*, vol. 115, no. 3, Aug. 1959. Aharonov and Bohm showed that an electromagnet could produce an effect in regions where its magnetic field was previously considered to be absent. The electromagnet produced a magnetic vector potential which exists where the classical magnetic field is zero.
5. http://en.wikipedia.org/wiki/Noether%27s_theorem
6. *The Coming Energy Revolution*, Jeane Manning, Avery Publishing Group, New York, 1996, page 55.
7. *Paramagnetism:Rediscovering Nature's Secret Force of Growth*, Philip S Callahan, Acres USA, 1995.
8. See http://www.landandspirit.net/html/geopathic_stress.html
9. *The Orgone Accumulator Handbook*, James DeMeo, Natural Energy Works, Ashland, OR. 2nd ed. 1999. DeMeo cautions against the use of aluminium. See pages 43, 61, 64.

Chapter 17   Applications of Subtle Energies

1. Borland C and Landrith G III, 'Improved quality of city life through the Transcendental Meditiation program', *Scientific Research on the Transcendental Meditiaton Program: Collected Papers*, Vol. 1, (pages 639-648) MERU Press Germany, 1976. also Dillbeck M C, Landrith G III and Orme-Johnson D W, 'The Transcendental Meditation program and crime rate changes in a

sample of forty eight cities', *Journal of Crime and Justice* Vol. 4, pages 25-45. also Bleick C R and Abrams A I, ' The Transcendental Meditation program and criminal recidivism in California', *Journal of Criminal Justice,* 15 (1987) pages 211-230.

2. *The Holographic Universe,* Michael Talbot, HarperCollins, London 1996 page 14 et seq.
3. Ibid. page 13. Rats were trained to run a maze and then parts of their brains were removed. Even though massive amounts of their brains were gone, they still retained the memory of how to find the way out of the maze.
4. *The Field,* Lynne McTaggart, Element, London 2001, page 124.
5. *The Body Electric,* Robert O. Becker and Gary Selden, William Morrow, New York, 1985, page 267. Becker's research on test organisms using Kirlian photography showed that the effect was produced by the water vapour in the organism and remained the same if the organism died so long as its water content remained the same. However since water may hold memories it would be unwise to conclude that life force was not being detected.
6. See http://www.kirlian.org/kirlian.htm
7. *Harry Oldfield's Invisible Universe*, Jane and Grant Solomon, Thorsons, London 1998, also www.electrocrystal.com
8. The piezoelectric effect occurs in certain crystals which produce a movement when an electric charge is applied. The effect also works in reverse where the crystals produce a voltage in response to an applied mechanical force.
9. *Hands of Light: A Guide to Healing Through the Human Energy Field*, Barbara Ann Brennan, Bantam 1987.
10. *The Rainbow in your Hands,* Albert Roy Davis and Walter C Rawls, Exposition Press, Smithtown Press N Y, 1976.
11. *Elegant Empowerment,* Peggy Phoenix Dubro and David P Lapierre, Platinum Publishing House 2002, page 127.
12. *Tuning the Diamonds*, Susan Joy Rennison, Joyfire Publishing, Staffordshire 2006. Page 174.

Chapter 18   The Power of Intention

1. *The Relaxation Response*, Herbert Benson MD, HarperCollins, New York, 2000

2. *Elegant Empowerment,* Peggy Phoenix Dubro and David P Lapierre, Platinum Publishing House 2002, page 106.
3. Hagelin J, 'Is Consciousness the Unified Field?', *Modern Science and Vedic Science*, Vol. 1 No.1 January 1987, pages 29-87, MUM Fairfield, Iowa
4. *Dialogues with Scientists and Sages; The Search for Unity,* Renée Weber, Routledge & Kegan Paul, London 1986 page 86.
5. *The Physiology of Consciousness*, Robert Keith Wallace, MIU Press, Fairfeld, Iowa 1993, page 239.
6. *The Tipping Point,* Malcolm Gladwell, Abacus, London, 2001.
7. *Self Hypnosis: A Safe Self-Help Guide,* Valerie Austin, Thorsons, London, 1994 page 24
8. Ibid.
9. Orme-Johnson D, "The Cosmic Psyche-An Introduction to Maharishi's Vedic Psychology: The Fulfillment of Modern Psychology". *Modern Science and Vedic Science*, Vol. 2, No. 2, Summer 1988, pages 113-163, MUM Fairfield, Iowa
10. *The Christian's Dilemma: A guide to the new spirituality*, Kieran Comerford, Elo Publications, Dublin, 2002, pages -6, 34, 81.
11. *The Intention Experiment,* Lynne McTaggart, HarperElement, London 2007.

Chapter 19   Not How But Whether

1. *The Seat of the Soul*, Gary Zukav, Rider, London 1990
2. Davenas E et al, 'Human basophil degranulation triggered by very dilute antiserum against IgE', *Nature* Vol. 333 (1988) pages 816-818.
3. *The Memory of Water: Homeopathy and the Battle of Ideas in the New Science*, Michel Schiff, Thorsons, London 1995.
4. 'Thanks for the Memory: Experiments have backed what was once a scientific 'heresy' says Lionel Milgrom', *The Guardian,* March 15, 2001
   http://www.guardian.co.uk/Archive/Article/0,4273,4152521,00.ht m
5. See: www.digibio.com
6. Horrobin David F, 'Something Rotten at the Core of Science', *Trends in Pharmacological Sciences*, 22, no 2, February 2001.

Also, Rothwell P M, Martyn C N 'Reproducibility of peer review in clinical neuroscience. Is agreement between reviewers any greater than would be expected by chance alone?' *Brain* 2000 Sep 123 (Pt 9):1964-9

7. *Paramagnetism:Rediscovering Nature's Secret Force of Growth*, Philip S Callahan, Acres USA, 1995. and www.orgonelab.org/energyblankets.htm

Chapter 20   Creating a New Environment

1. *The Seven Spiritual Laws of Success*, Deepak Chopra, Amber-Allen Publishing, San Rafeael, CA, 1993, page 11
2. *The Power of Now,* Eckhart Tolle, Hodder & Stoughton, London, 2001.
3. *A New Earth,* Eckhart Tolle, Michael Joseph, London 2005, page 27. The correct quote is "an optical delusion of consciousness". (Einstein 1954). Maybe this is a translation from German and Eckhart read it in the original.
4. *My Stroke of Insight: A Brain Scientist's Personal Journey,* Jill Bolte Taylor, Penguin, 2008. Also http://drjilltaylor.com/
5. *Be Here Now*, Ram Dass, Hanuman Foundation, 1978, further details at www.ramdasstapes.org
6. *Weaving the Tapestry of Our Dreams*, Ram Dass, Keynote address to NLP Conference, Denver CO, Sept. 16, 1994, www.ramdasstapes.org
7. Bleick C R and Abrams A I, ' The Transcendental Meditation program and criminal recidivism in California', *Journal of Criminal Justice,* 15 (1987) pages 211-230.
8. See http://www.amasci.com/freenrg/tors/tors3.html
9. *The Orgone Accumulator Handbook*, James DeMeo, Natural Energy Works, Ashland, OR. 2nd ed. 1999.
10. See www.rexresearch.com/torsion/torsion1.htm In this article by Nachalov and Sokolov it is stated that aluminium can sometimes screen torsion fields.
11. *Energy Medicine: The Scientific Basis*, James L. Oschmann, Elsevier 2000.

Appendix A

1. See Chapter 4 note 2 above.
2. See Chapter 18.
3. *The Christian's Dilemma: A guide to the new spirituality*, Kieran Comerford, Elo Publications, Dublin, 2002, page 85. Also, chapter 9 of the same book gives an overview of meditation.
4. *The Miracle of Mindfulness: A Manual on Meditation,* Thich Nhat Hahn, Rider, London, 1987

Appendix B

1. *The Orgone Accumulator Handbook*, James DeMeo, Natural Energy Works, Ashland, OR. 2nd ed. 1999, pages 43, 61, 64.
2. See: http://www.orgonelab.org/energyblankets.htm
3. See: http://www.hpa.org.uk/radiation/understand/information_sheets/mobile_telephony/mobile_phones.htm
4. See: http://news.bbc.co.uk/2/hi/programmes/panorama/6674675.stm

Appendix C

1. *The Sense of Being Stared at and Other Aspects of Extended Mind*, Rupert Sheldrake, Hutchinson London, 2003.
2. W. Braud and M. Schlitz, "Psychokinetic Influence on Electro-Dermal Activity", *Journal of Parapsychology*, vol.47, 1988, pages 95-119 as quoted in *Miracles of Mind: Exploring Nonlocal Consciousness and Spiritual Healing,* Russell Targ and Jane Katra, New World Library, California, 1998, page 212.
3. *The Christian's Dilemma: A guide to the new spirituality*, Kieran Comerford, Elo Publications, Dublin, 2002, page 27
4. *The Field,* Lynne McTaggart, Element, London 2001, page 191

5. *Miracles of Mind: Exploring Nonlocal Consciousness and Spiritual Healing,* Russell Targ and Jane Katra, New World Library, California, 1998,

6. Harold Puthoff and Russell Targ, " A Perceptual Channel for Information Transfer over Kilometer Distances: Historical Perspective and Recent Research, Proceedings of IEEE, vol 64, no.3, March 1976, pages 329-354

7. McTaggart op. cit. page 199.

INDEX

www.ingramcontent.com/pod-product-compliance
Lightning Source LLC
Chambersburg PA
CBHW030004190526
45157CB00014B/418